7th International Conference on Mechanics and Mechatronics (ICMMR 2020)

Held online due to COVID-19

IOP Conference Series: Materials Science and Engineering
Volume 938

Berkeley, California, USA
26 - 28 June 2020

ISBN: 978-1-7138-3177-8
ISSN: 1757-8981

Printed from e-media with permission by:

Curran Associates, Inc.
57 Morehouse Lane
Red Hook, NY 12571

proceedings.com

Some format issues inherent in the e-media version may also appear in this print version.

Institute of Physics
Dirac House, Temple Back
Bristol BS1 6BE UK

Phone: 44 1 17 929 7481
Fax: 44 1 17 920 0979

techtracking@iop.org

Additional copies of this publication are available from:

Curran Associates, Inc.
57 Morehouse Lane
Red Hook, NY 12571 USA
Phone: 845-758-0400
Fax: 845-758-2633
Email: curran@proceedings.com
Web: www.proceedings.com

TABLE OF CONTENTS

IOP Conf. Series: Materials Science and Engineering **938** (2020) 011001 doi:10.1088/1757-899X/938/1/011001

Preface

Welcome to the 2020 7th International Conference on Mechanics and Mechatronics Research (ICMMR2020). This year, our annual Conference was planned to be held on the campus of the University of California, Berkeley, 27-29 June 2020. All technical sessions and meal functions was planned to be held in beautiful Bowles Hall, which is a registered historic building on campus. The weather in Berkeley should be wonderfully sunny and neither too hot nor too cold; perfect for strolls around campus and the San Francisco Bay Area. The conference offered an opportunity for researchers from around the world to exchange ideas and information in an idyllic setting in one of the great educational institutions of the world. A wide of array of keynote talks, oral presentations, posters sessions, and tours are planned. But, unfortunately, due to the spread of the COVID-19 virus around the world, we have to curry on our presentations online.

The conference program included the keynote and oral presentations from scientists working in similar areas to establish platforms for collaborative research projects. The time for each keynote speaker was 45 minutes, while for authors was 15 minutes each. All keynote, parallel sessions, discussions and other activities were offered in a streaming online. The conference covered the recent trends and achievements in the fields of mechanics and mechatronics research.

This year we have around twenty participants from USA, Russia, UAE, Indonesia, Jordan, South Korea and other countries, whom we wish fruitful discussions. ICMMR2020 created a customized online platform that was a virtual environment where the attendees could present the papers and participate in all the sessions that were organized in a full immersive experience. We certainly proved to be an informative, useful, and pleasant experience.

Prof. Dennis K. Lieu
Professor of Mechanical Engineering UC Berkeley
Host Chair, 2020 ICMMR

Committee

INTERNATIONAL ADVISORY COMMITTEE
Prof. Allison M. Okamura (IEEE Fellow), Stanford University, United States

CONFERENCE COMMITTEE CHAIR
Prof. Dennis Lieu, University of California, Berkeley, United States

PROGRAM COMMITTEE CHAIR
Prof. Dr. Jan Awrejcewicz, Lodz University of Technology, Poland
Marina Barulina, Institute of Precision Mechanics and Control, Russian Academy of Sciences, Russia

INTERNATIONAL TECHNICAL COMMITTEES
Prof. He Xu, Harbin Engineering University, China
Prof. Kwanho You, Sungkyunkwan University, South Korea
Prof. Mohamed bin Hussein, Universiti Technologi Malaysia, Malaysia
Assoc. Prof. Yi Jin, University of Science and Technology of China, China
Assoc. Prof. Dirman Hanafi, Universiti Tun Hussein Onn Malaysia, Malaysia
Assoc. Prof. Dr. Vyacheslav Petrenko,North-Caucasus Federal University, Russia
Assoc. Prof. Dr. Maria Lapina, North-Caucasus Federal University, Russia
Asst. Prof. Abdel-Hameed Badawy, New Mexico State University, USA
Dr. Zahid Akhtar, University of Memphis, USA
Dr. Hsien-I Lin, National Taipei University of Technology, Taiwan
Dr. Lihong Zheng, Charles Sturt University, Australia
Dr. Adrian Kampa, Silesian University of Technology, Poland
Dr. Grzegorz Ilewicz, University of Bielsko-Biala, Poland
Ibragim Suleimenov, Almaty University of Power Engineering and Telecommunications, Kazakhstan
Akhat Bakirov, Almaty University of Power Engineering and Telecommunications, Kazakhstan
Sebastian Heil, Chemnitz Technical University, Germany

IOP Conf. Series: Materials Science and Engineering **938** (2020) 011003 doi:10.1088/1757-899X/938/1/011003

Virtual Disclaimer

As the coronavirus (COVID-19) spreads worldwide and taking into account the various circumstances of the participants, this year the conference was held in online conference. We were creating a customized "ICMMR 2020 online platform" that was a virtual environment where the attendees could present the papers and could participate in all the sessions that were organized in a full immersive experience. The access to the ICMMR online platform was offered to all the registered attendees. The virtual conference was held in Berkeley, USA during June 27-29, 2020. The time for each keynote speaker was 45 minutes, while for author was 15 minutes each. All keynote, parallel sessions, discussions and other activities were offered in a streaming online. This year we have around twenty participants from USA, Russia, UAE, Indonesia, Jordan, South Korea and other countries, whom we wish fruitful discussions.

IOP Conf. Series: Materials Science and Engineering **938** (2020) 011004 doi:10.1088/1757-899X/938/1/011004

Peer review declaration

All papers published in this volume of *IOP Conference Series: Materials Science and Engineering* have been peer reviewed through processes administered by the Editors. Reviews were conducted by expert referees to the professional and scientific standards expected of a proceedings journal published by IOP Publishing.

- **Type of peer review:** Double-blind
- **Conference submission management system:** Iconf submission system
- **Number of submissions received:** 17
- **Number of submissions sent for review:** 17
- **Number of submissions accepted:** 10
- **Acceptance Rate (Number of Submissions Accepted / Number of Submissions Received X 100):** 58.82%
- **Average number of reviews per paper:** 3
- **Total number of reviewers involved:** 6
- **Any additional info on review process:**

  1. Preliminary Review
  2. The papers passed the first review will be reviewed again from the following aspects: Originality, Innovation, Technical Soundness, Applicability, Clarity of presentation and Relevance

- **Contact person for queries:** Jian Yang (26199636@cqu.edu.cn)

IOP Conf. Series: Materials Science and Engineering **938** (2020) 012001 doi:10.1088/1757-899X/938/1/012001

Multifunctional Carbon Fiber Composite with Improved Electromagnetic Interference Shielding and Interlaminar Fracture Toughness Characteristics

Nisrin R Abdelal

Associate professor, Department of Mechanical Engineering, Jordan University of Science and Technology, Irbid 22110 Jordan

nrabdelal@just.edu.jo

Abstract. In this study, multifunctional carbon fiber-epoxy composites with improved electromagnetic interference shielding and interlaminar fracture toughness were manufactured. The interest in manufacturing multifunctional composites has been driven by the need for materials that simultaneously perform structural and non-structural functions. Unidirectional carbon fiber-epoxy composite laminates stitched with Dyneema13, Copper, and Kevlar13 were prepared using resin infusion process. Representative specimens from the unstitched and the stitched composites were tested using rectangular waveguide and Mode I interlaminar fracture tests. The electromagnetic shielding effectiveness experimental results showed that stitched composites exhibited improved shielding effectiveness compared to unstitched composites. For example, composite stitched with copper showed the highest increase of 102.2% in shielding effectiveness whereas composite stitched with Dyneema13 showed the smallest increase of 75.4% in shielding effectiveness compared to the unstitched. Results revealed that the dominant effect of shielding in these composites is absorption. The Mode I experimental results showed that composite stitched with Kevlar13 exhibited the highest crack initiation interlaminar fracture toughness ($G_{IC\text{-}initiation}$), whereas composite stitched with Dyneema13 exhibited the highest maximum crack propagation interlaminar fracture toughness ($G_{IC\text{-}maximum}$). The experimental results comply with the goal of the study to produce multifunctional composite with improvement in the mechanical and electromagnetic properties.

1. Introduction

Nowadays, manufacturing multifunctional materials having simultaneous improvements in the mechanical and the electrical properties is a major area of interest that is driven by the need for materials and systems that simultaneously perform structural and non-structural functions[1]. Composites by nature are multifunctional materials. For example, fiber reinforced polymer (FRP) composites have excellent structural properties such as high specific strength and high stiffness. However, FRPs have poor electrical properties due to the dielectric nature of the polymer matrix. Thus, incorporating conductive nanofillers in the matrix such as carbon nanotubes will improve their electrical and electromagnetic shielding properties [2]–[6]. The superior structural properties of FRP composites made them the best candidate for manufacturing aircraft structures. Whereas the non-structural properties such as the electromagnetic interference shielding (EMI-SE), optical transparency, and thermal conductivity made them perfect candidates in applications such as in stealth applications, electronics, piezoelectric materials.

Despite the remarkable properties of FRPs, they are susceptible to delamination. Delamination is a serious failure mode in FRPs. It is a separation of the FRP layers caused by out-of-plane loads. Several methods have been developed to strengthen FRPs through- the-thickness and resist delamination such as stitching, z-pinning, 3D weaving and incorporating nanofillers in the polymer matrix[7]–[9]. One effective method in alleviating delamination is stitching the composite fabrics through the thickness using high strength threads such as Kevlar and carbon. Stitching is an excellent method in resisting delamination and improving the interlaminar fracture toughness (IFT) of FRP composites [10]–[12]. Therefore, stitching was the chosen method in this study to investigate its ability in resisting delamination in carbon fiber reinforced polymer composite using non-traditional types of thread (conductive thread copper and nonconductive Dyneema) and compare it with the traditional threads (Kevlar). Embedding conductive materials as threads, layers or particles in the polymer matrix was used in studies to improve the electrical and electromagnetic properties of the FRP composites[13]–[15]. In this direction, the conductive thread (copper) was chosen for stitching to investigate its ability in improving the EMI-SE properties and the IFT of the composite. For comparison purposes, nonconductive threads (Dyneema and Kevlar) were chosen to investigate their abilities in improving the EMI properties of the composite in addition to the anticipated improvement in the IFT. Unstitched laminate was manufactured as pristine (control) laminate.

2. Methodology

2.1. Materials
The carbon fiber composite laminates were fabricated using dry TORAY T-700- 300gsm-12K-unidirectional (UD) carbon fibers, CE-R3501 vacuum infusion epoxy and CE-H5000-01 curing agent, all supplied from Composite Envisions-USA. The carbon fiber has an electrical resistivity of $1.6X10^{-3}$ Ω.cm and 4900 MPa tensile strength. Three types of stitching threads were used: Kevlar 13, Dyneema13 and Copper. The Kevlar thread has 0.3mm diameter and 13Kg maximum tensile load, Dyneema thread has 0.3mm diameter 13Kg maximum tensile load, and copper has 0.25mm diameter, 5.3Kg maximum tensile load, and an electrical resistivity of 8 μOhm.cm.

2.2. Composite Laminate Fabrication
One unstitched laminate as (pristine) control composite and three stitched laminates were manufactured. Each laminate was made of 10 plies of UD carbon fabrics arranged in the 0° orientation $[0°]_{10}$. In the stitched laminates, Kevlar13, Dyneema13 and Copper were used to stitch the dry carbon fabrics through the thickness as described in [16]. All laminates were manufactured using the vacuum assisted resin transfer molding (VARTM) method following the procedure described in [17]. A total of four laminates (three stitched and one unstitched) were manufactured to study the effect of stitching with different types of threads on Mode I-IFT and EMI-SE of the composites.

2.3. Specimen preparation
Specimens for the Mode I-IFT were cut with dimensions in accordance with the ASTM standard D5528[18] were the unidirectional (UD) fibers were oriented along the length (L). specimens were prepared for the Mode I test as described in [17]. On the other side, specimens for the electromagnetic interference shielding measurements were cut with the dimensions 3cm X 3cm.

2.4. Electromagnetic interference shielding effectiveness measurements
The specimens were tested according to the setup described in [16]. At least five specimens from each laminate were tested for (EMI-SE) properties. Measurements were conducted using E5071CENA series 300 kHz-20GHz Network Analyzer connected to WR90 rectangular waveguide (SIVERS lab – Philips PM-7328-X) via coaxial cables at the X-band high frequency range (8–12 GHz). Each specimen was inserted between the waveguide's parts and the scattering parameters S-parameters (S_{11}, S_{12}, S_{21}, S_{22}) were recorded. Each specimen was tested in two orientations, the axial direction where the UD fibers of the specimen were aligned parallel to the long side of the waveguide, and the normal direction where the UD fibers were aligned parallel to the short side of the waveguide. Using the S-

parameters, the total EMI-SE, the absorption loss (SE_A), the reflection loss (SE_R) and the electrical conductivity of each specimen were calculated using equations (1-5) [19]–[22]:

$$EMI\ SE = SE = 10\ log\frac{1}{|S_{12}|^2} = 10\ log\frac{1}{|S_{21}|^2} \quad (1)$$

$$EMI\ SE = SE_R + SE_A \quad (2)$$

$$SE_A = 10\log\left(\frac{1-|S_{11}|^2}{|S_{12}|^2}\right) \quad (3)$$

$$SE_R = 10\log\left(\frac{1}{1-|S_{11}|^2}\right) \quad (4)$$

$$\sigma^2 = 2 \times \left(\frac{SE_A}{8.7d}\right)^2 \times 10^{\left(\frac{SE_R - 39.5}{10}\right)} \quad (5)$$

Where $|S_{ij}|^2$ is defined as the power ratio; S_{12} is the ratio between the transmitted power through the shield to the incident power and S_{11} is the ratio between the reflected power from the shield to the incident power, and d is the sample's thickness.

2.5. Mode I – Interlaminar fracture test

The tensile properties of the stitching threads (Kevlar13, Dyneema13 and copper) were determined in [1]. Mode I tests were conducted following the ASTM D5528 standard using a universal testing machine (jinan-WDW20) equipped with 2KN load cell. The tests were run in a displacement controlled mode with a 2mm/min crosshead speed. Seven specimens from each laminate were tested to calculate mode I-IFT (G_{IC}) at multiple crack lengths (a) following the procedure described in [17]. The Mode I interlaminar fracture toughness (energy release rate) (G_{IC}) was calculated using the Modified Beam Theory with the correction factor Δ (MBT-Δ) using equation (6) shown below [18].

$$G_{IC} = \frac{3P\delta}{2b(a+|\Delta|)} \quad (6)$$

Where P is the applied load (N), δ is the opening displacement (mm), b is the width of the sample (mm), (a) is the crack (delamination) length (mm), and Δ is a correction for crack tip rotation.

3. Results and Discussion

The interlaminar fracture toughness results in section 3.1 were previously reported in reference [17]. However, in this study the main theme was to emphasize on the fact that the author was able to manufacture a multifunctional composite with improved electromagnetic interference shielding and electrical conductivity in addition to the previously reported improved mechanical properties. In this study, Dyneema13 was selected because the composite stitched with it showed the highest IFT in [17] and its EMI-SE properties were not investigated before. Kevlar13 was selected because its EMI-SE properties were not studied before, and copper was selected as a type of conductive thread.

3.1. Mode I interlaminar fracture toughness of unstitched and stitched composites

Figure 1 summarizes all Mode I results of both unstitched and stitched specimens. The mean of all tested specimens' initial, minimum, maximum, final and average values of G_{IC} were calculated and plotted as a bar. The figure shows that the initiation fracture toughness ($G_{IC\text{-}initial}$) of the specimens stitched with copper (1.14 kJ/m^2) and Dyneema13 (1.43 kJ/m^2) corresponds to 100% and 150% increase respectively compared to the unstitched composite (0.57 kJ/m^2). However, specimens stitched with Kevlar13 (2.06 kJ/m^2) showed the highest increase in ($G_{IC\text{-}initial}$) 260% compared to the unstitched. The maximum fracture toughness ($G_{IC\text{-}max}$) of the specimens stitched with Dyneema showed extremely high increase of 986% compared to those stitched with Kevlar (increase of 291%) and copper (increase of 375%) compared to the unstitched. Note that the highest initial value was for composite stitched with Kevlar whereas the highest maximum value was for the composite stitched with Dyneema. The average fracture toughness ($G_{IC\text{-}average}$) follows a trend similar to ($G_{IC\text{-}max}$).This lead to select these composites to further analyze their electromagnetic shielding characteristics to investigate the possibility of classifying them as multifunctional composites.

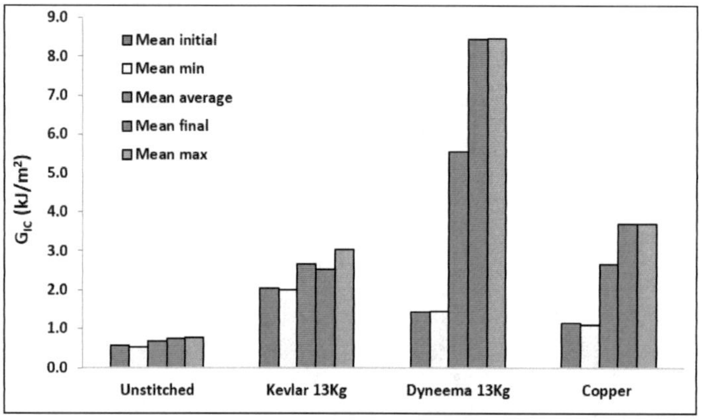

Figure 1. Summary of the Mode I-DCB results: The mean values of the initiation, minimum, maximum, final and average interlaminar fracture toughness (G_{IC}-initiation, G_{IC}-minimum, G_{IC}-final, G_{IC}-maximum and G_{IC}-average) with the standard deviation error bars.

3.2. Electromagnetic interference shielding

For each tested sample from the four laminates, the overall electromagnetic shielding effectiveness (SE), the absorption loss (SE_A) and the reflection loss (SE_R) were calculated in the 8-12GHz frequency range with frequency increment of 0.02 GHz. Figure 2 shows the SE versus frequency of unstitched and stitched composites. It is clear that composites stitched with conductive (Cu) and nonconductive (Kevlar13 and Dyneema13) threads exhibit remarkable increase in the SE compared to the unstitched composite. However, composite stitched with copper exhibits the highest shielding effectiveness values amongst the stitched composites in the 9-12 GHz range.

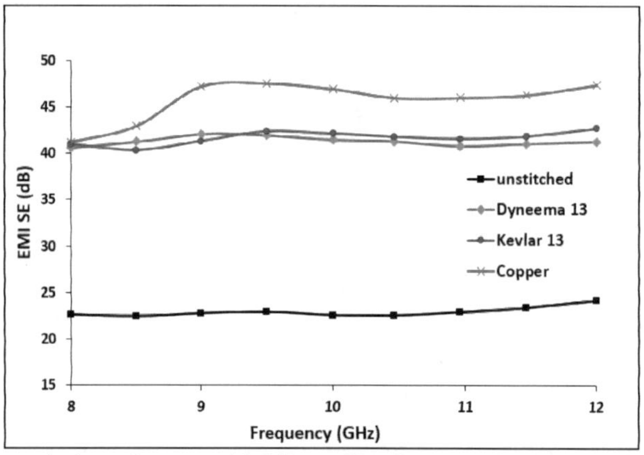

Figure 2. Electromagnetic interference shielding effectiveness (EMI-SE) of samples representing unstitched and stitched composites in the axial direction.

Figure 3.A&B show the absorption Loss SE_A and the reflection loss SE_R versus frequency of the unstitched and stitched composites. Figure 3.A shows that the absorption Loss SE_A almost follows the same trend as the SE where it increases with increasing the frequency whereas Figure 3.B shows that the reflection loss SE_R decreases with increasing the frequency. This is because the material's absorptivity of the electromagnetic waves increases with increasing the frequency and its reflectivity decreases with increasing the frequency.

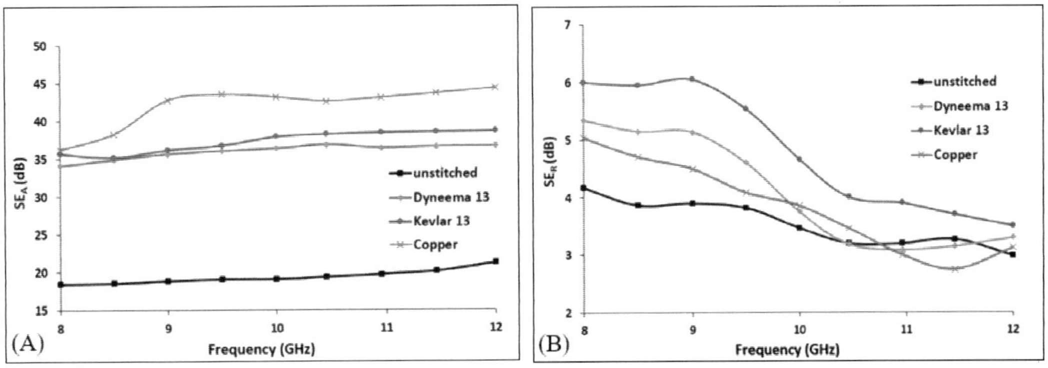

Figure 3. (A) Absorption Loss SE_A and (B) reflection loss SE_R of samples representing unstitched and stitched composites in the axial direction

The SE results of all the tested specimens from the four laminates are summarized in Table 1. For each shielding effectiveness term (SE, SE_A, SE_R), the average value of each tested specimen was calculated over the 8-12 GHz range, then the average SE of the five testes specimens was calculated and included in the table. It is clear that stitched composites regardless the type of stitching thread have higher SE values in the axial direction compared to the unstitched with a maximum increase of 102.2% when stitching with copper and smallest increase of 75.4% when stitching with Dyneema. It is believed based on the findings from previous studies that this behavior is due to the increase in the composite conductivity in the three dimensions (along the fibers, transverse and through the thickness) because stitching compress the fibers together in all three directions which lead to creation of conductive channels for the electromagnetic waves to flow. A more detailed explanation was introduced in [16].

Table 1. Average values of: electromagnetic shielding effectiveness (EMI-SE) in both axial and normal orientations, absorption loss (SE_A), reflection loss (SE_R) and the conductivity

	SE-axial (dB)	SE-Normal (dB)	SE_A (dB)	SE_R (dB)	σ (S/m)
Unstitched	22.8	63.5	19.5	3.5	13
Copper	46.1	58	42.1	3.7	25.5
Dyneema13	40	50	35.7	3.9	27.5
Kevlar13	41.5	69.7	36.9	4.5	31.14

In the case of composites in the normal direction, there is no effect of stitching on the SE values. This is because the carbon fibers are aligned with the polarization direction of electric field in the TE mode (fibers parallel to the short wall of the waveguide) which increases the overall conductivity of the composite. In fact stitching with Dyneema and Kevlar lead to a slightly smaller SE in the normal direction compared to the unstitched. This is because Dyneema and Kevlar have larger thread diameter with multi filaments compared to the single filament copper, which led to the creation of the nonconductive resin pockets between the fibers. These resin pockets lower the conductivity along the fiber directions and block the flow of the electrical charges. Finally, Table 1 shows the absorption loss (SE_A) for each type of composite is higher than the reflection loss (SE_R). This means the dominant effect of shielding in these composites is absorption. The results achieved were in line with previous research studies. It is also clear that SE_A is higher for stitched composites compared to the unstitched with a maximum increase of 116% when stitching with copper and smallest increase of 83.1% when stitching with Dyneema. This is because the electromagnetic wave's wavelength decreases at high frequency ranges (X-band) and becomes closer to the fiber size; therefore, it is easier to be absorbed than reflected by the composite.

4. Conclusions

This study experimentally investigated the possibility of manufacturing a multifunctional composite with improved mechanical and EMI-SE properties. Composites stitched through the thickness with conductive (copper) and nonconductive (Kevlar and Dyneema) threads with a pristine composite to were prepared using resin infusion process. EMI-SE test results showed that stitched composites have higher shielding effectiveness (SE) compared to the unstitched composites. The maximum SE was achieved by stitching with copper whereas the least was achieved by stitching with Dyneema13. Results showed that the dominant effect of shielding is absorption. Mode I-IFT tests revealed that composite stitched with Dyneema13 has the highest maximum IFT whereas composite stitched with Kevlar 13 has the highest initiation IFT. Goal of the study was fulfilled by manufacturing composites with improved resistance to delamination and EMI-SE using conductive and non-conductive threads.

5. References

[1] J. Narayana and R. Gupta, *Mater. Today Proc.*, vol. 5, no. 2, pp. 5580–5590, 2018.
[2] H. Chang, Y.-M. Yeh, and K.-D. Huang, *Mater. Trans.*, vol. 51, no. 6, pp. 1145–1149, 2010.
[3] S. Yang, K. Lozano, A. Lomeli, H. D. Foltz, and R. Jones, *Compos. Part A Appl. Sci. Manuf.*, vol. 36, no. 5, pp. 691–697, 2005.
[4] E. J. Garcia, B. L. Wardle, A. John Hart, and N. Yamamoto, *Compos. Sci. Technol.*, vol. 68, no. 9, pp. 2034–2041, 2008.
[5] T D Jagannatha and G Harish, *Int. J. Mech. Eng. Robot. Res.*, vol. 4, no. 2, 2015.
[6] C. K. Eric and Z. Yu, vol. 6, no. 1, pp. 11–15, 2017.
[7] M. V. Hosur, U. K. Vaidya, C. Ulven, and S. Jeelani, *Compos. Struct.*, vol. 64, no. 3–4, pp. 455–466, 2004.
[8] S. U. Khan and J. K. Kim, *Int. J. Aeronaut. Sp. Sci.*, vol. 12, no. 2, pp. 115–133, 2011.
[9] A. P. Mouritz, *Compos. Part A Appl. Sci. Manuf.*, vol. 38, no. 12, pp. 2383–2397, 2007.
[10] R. F. Creed, no. March, p. 84, 1993.
[11] M. R. Pankow, A. M. Waas, and L. Nicolais, *Wiley Encycl. Compos.*, no. July 2012, pp. 1–7, 2011.
[12] K. Tan, N. Watanabe, and Y. Iwahori, *J. Reinf. Plast. Compos.*, vol. 30, pp. 99–109, 2011.
[13] K. B. Cheng, S. Ramakrishna, and K. C. Lee, *Compos. Part A Appl. Sci. Manuf.*, vol. 31, no. 10, pp. 1039–1045, 2000.
[14] R. Perumalraj and K. S. Narayanan, *J. Reinf. Plast. Compos.*, vol. 33, no. 11, pp. 1000–1016, 2014.
[15] F. Ceken, G. Pamuk, O. Kayacan, A. Ozkurt, and Ş. S. Ugurlu, *J. Eng. Fiber. Fabr.*, vol. 7, no. 4, pp. 81–87, 2012.
[16] N. Abdelal, *J. Ind. Text.*, vol. 49, no. 6, pp. 773–790, 2018.
[17] N. R. Abdelal and S. L. Donaldson, *Polym. Compos.*, vol. 40, no. S2, pp. E1252–E1262, 2018.
[18] a Standard, *Am. Soc. Test. Mater.*, pp. 1–13, 2014.
[19] M. H. Al-Saleh, *Synth. Met.*, vol. 205, pp. 78–84, 2015.
[20] M. H. Al-Saleh, *J. Phys. D. Appl. Phys.*, vol. 49, no. 19, p. 195302, 2016.
[21] N. Li, Y. Huang, F. Du, X. He, X. Lin, and H. Gao, *Nano Lett.*, vol. 6, pp. 1–5, 2006.
[22] Z. Liu *et al.*, *Carbon N. Y.*, vol. 45, no. 4, pp. 821–827, 2007.

Acknowledgments

This work was financially supported by the Deanship of Research - Jordan University of Science and Technology, Irbid– Jordan, Research grants with the numbers (184/2015) and (/2016)

The Innovation of Using 3D Printing Technology in Mechanical and Manufacturing Engineering

Waleed Ahmed[1], **Essam Zaneldin**[2]

[1]Engineering Requirements Unit, College of Engineering, UAE University, Al Ain, UAE

[2]Civil and Environmental Engineering Department, College of Engineering, UAE University, Al Ain, UAE

Corresponding author: w.ahmed@uaeu.ac.ae

Abstract. One of the main challenges that the educator faces in integrating 3D printing technology in a traditional engineering course in the capability of introducing and merging the 3d printing technology at the right educational material. The paper aims to present the challenges and the achievements of integrating 3D printing technology in teaching design and manufacturing courses in the mechanical engineering department to enhance the students' skills and engagement with the course material. One of the course experiments was nominated to introduce the 3D printing technology, where the students were asked to use the systematic design method as well as design matrix procedure to explore different options to come up with a solution to hold a mechanical component for quality control task. Besides, the students had to design an engineering measurement procedure to achieve the assigned mission of the experiment, where the mechanical part was described to present the main dimensions needed for the measurement sequence. Different groups submitted innovative ideas that were evaluated initially and discussed individually before submitting the final design and the holder prototype. The students presented a functional 3D printed prototype with a full report that clarifies the engineering procedure adopted to achieve the goal of the experiment. The innovation in the addressed solution revealed the importance of 3D printing technology to enhance the students' skill and engagement with the course material and help them to encounter challenges to come up with creative engineering solutions.

1. Introduction

Additive manufacturing, or what is named 3D printing technology has a significant contribution to the world in many directions because of the remarkable advantages to the industrial sector. It has a substantial impact on other fields [1,2]. Moreover, this revolution has been extended to other applications, like in prototyping, simulations, and failure mechanism [3,4], whereas the aerospace sector is one of the promising areas due to the vast and numerous applications that could be adopted of using 3D printing technology [5]. However, the 3D printing technology started being embraced in schools [6] for the learning purposes, as well as implemented in different university levels [7]. At the university level, vocational, technical institutes, and universities throughout the world have long been offering courses related to technical drawing, engineering graphics, and computer-aided design (CAD) at various levels of different programs. 3D printed prototypes are very common in the arts, animation, and fashion studies as well. 3D printing can provide a simple tool for illustration and visualization for students allowing them to grasp the concepts of their coursework [8] quickly. Not only in the design and prototyping courses where students print static prototypes but also in other classes involving the complex movement of certain parts such as compressors and linkages where students suffer from visualizing the action at different instances. Hence, a 3D printed prototype with details involving these parts can solve this problem by providing a real-time simulation of the actual movement. Moreover, some courses, such as structural design and mechanics of materials, involve specific testing of materials and structures subjected to different loads. 3D printing can translate the problems discussed on the board by the instructor or even presented in the student's textbooks into a live model that can be experienced by the students allowing them to see the behavior of these structures under specific loads. The printed model can even be tested in the laboratory for tension or compression, bending, buckling, or under a sophisticated type of mechanical loading. This work will be focused to discuss the application of 3D printing in the following courses:

A. Mechanics of Materials
B. Manufacturing processes

2. Innovation of Using 3D Printing Applications in Mechanics of Materials: Case Study

In general, students sometimes face a particular sort of difficulty in perceiving the mechanics of material cases, where understanding the actual behavior of the structure is not easy. Therefore, by translating the case study into physical structure would enhance the student's grasp to overcome the imagination problem. Besides, engaging the students in assembling and disassembling the prototype will be a significant advantage to minimize the possible deficiency, uniquely combining both the 3D printed prototype with the actual bolts, nuts, and washers used in the structure. A problem from the mechanics of material textbook [9] was selected as a case study in the course to be considered in this practice, where it is anticipated many lessons will be achieved through this case. Uniquely it blends many theoretical backgrounds like dealing with free body diagram, statics analysis, trigonometry, tensile and compressive stresses, types of the supports, etc. There is much commercial software available to be used for design purposes, like Autocad, Solid Works, CATIA, Fusion 360, and the online Autodesk Free software Tinkercad. Besides, STL, OBJ, AMF, and 3MF are the four most common 3D printer file formats, but there are many others [10]. The case was designed and printed by the instructor of the course [11]. A group of students that finished the geometric modeling course MECH 315 was nominated to develop and to build by 3D printing the case under the instructor's supervision. The theoretical analysis of the case is depicted by the prototype made, it term of the geometry and the shape matching, and this helped the students to perceive the solution since it involves trigonometric analysis to find the angle of the rotation that will be used to calculate the extensions of the axial members at points A and C. Figure 1 illustrates a comparison between the theoretical case nominated from the mechanics of materials textbook and the 3D printed prototypes that show the similarity in the design.

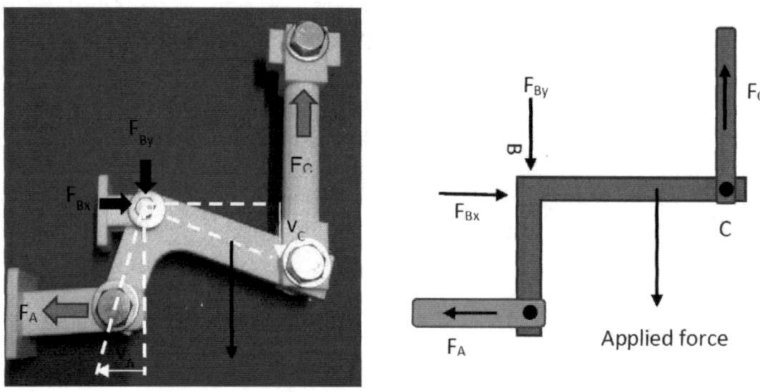

Figure 1. Comparison between the hypothetical case and the 3D printed prototype.

Where:
F_C: Axial tensile force caused due to elongation of the vertical member connected with joint C.
F_A: Axial compressive force caused due to the shortening of the horizontal member connected with joint A.
F_{Bx}: Horizontal component of the force reaction in joint B.
F_{By}: Vertical component of the force reaction in joint B.
v_C: Vertical displacement of joint C.
v_A: Horizontal movement of joint A.

Structural components that are subjected only to tension or compression are known as axially loaded members. Bars with straight longitudinal axes are the most common type, although cables and coil springs also carry axial loads. Examples of axially loaded bars are truss members, connecting rods in engines, spokes in bicycle wheels, columns in buildings, and struts in aircraft engine mount [12]. Axially loaded bars elongate under tensile loads and shorten under compressive loads. Unfortunately, most of the two-dimensional drawings don't represent the real case studied, and this may cause difficulties to the students to grasp the idea behind the practice and sometimes causes confusion and leads eventually to a wrong solution. As an example, a hollow, brass circular pipe ABC supports a load P1 acting at the top [12]. A second load P_2 is uniformly distributed around the cap plate at B. The diameters and thicknesses of the upper and lower parts of the pipe are d_{AB}, t_{AB}, d_{BC}, and t_{BC}. In contrast, the modulus of elasticity is E. It is observed from this case that the geometry details are not apparent to the students, especially if they did not take the engineering drawing yet. However, a 3D printed object that represents the case was constructive to overcome the confusion. On the other hand, some analyzed bars consisted of prismatic segments that may have different axial forces, different dimensions, and different materials. As an example, a non-prismatic cantilever bar has an internal hole of d/2 from 0 to x, so the net area of the cross-section for Segment 1 is (3/4)*A, as shown in Figure 2-a. The bar is subjected to an axial tensile load P at x, and another axial compressive load P/2 is applied at the end of the bar. The bar has a modulus of elasticity E that is constant along with the member. The problem was again to identify the geometry details, especially the depth of the bar. A designed model that represents the actual geometry was printed using 3D printed and used in teaching that was reflected positively on the students' grasp. This is illustrated in Figure 2-b.

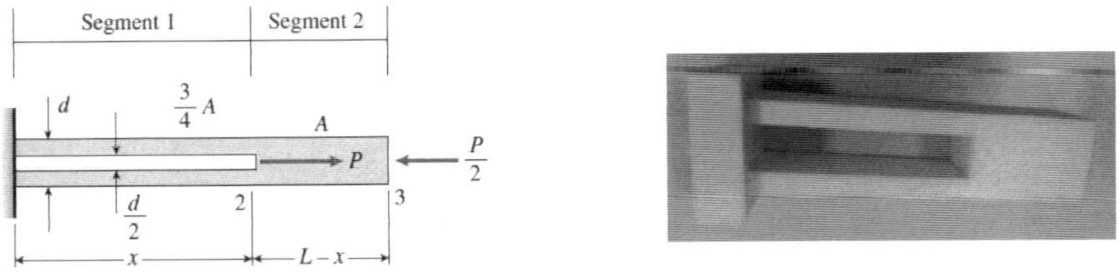

(a) (b)

Figure 2. A non-prismatic axially loaded cantilever bar that contains a hole (a), 3D printed design for the bar (b).

A composite structure system that contains embedded elements, like reinforced concrete and other applications, is another challenge that is usually faced in teaching the mechanics of materials. As an example, a cylinder and a tube are compressed between the rigid plates of a testing machine by compressive forces P [12]. The steel cylinder has cross-sectional area As and modulus of elasticity Es, the copper tube has area Ac and modulus Ec, and both parts have length L. In order to solve the problem, the student must imagine the free body diagram of the composite structure, which is a difficult task since the imagination level varies from a student to another. Therefore, printing a 3D model by the 3D printer, this will diminish the imaginary weakness of some students. Besides, the students will have better images to deal with the deformation of the composite system, since they are going to test the experience of assembling and disassembling of the composite structure themselves. Accordingly, at the next stage of the solution, they will have to imagine how both elements will have the same axial shortening. This structure is classified as statically indeterminate since the numbers of the unknowns are more than the equations available. To analyze such arrangements, we must supplement the equilibrium equations with additional equations pertaining to the displacements of the structure. A knuckle joint is used to connect two rods that are under the action of tensile loads (Figure 3). However, if the joint is guided, the rods may support a compressive load. A knuckle joint may be readily disconnected for adjustments or repairs. It has many applications such as it may be found in the link of a cycle chain, tie rod joint for roof truss, valve rod joint with the eccentric rod, pump rod joint, tension link in bridge structure and lever and rod connections of various types [13]. The knuckle joint example is a typical case that is used in the machine design course since it has many failure modes that students need to understand them one by one. Due to the complexity of the knuckle joint, students always encounter difficulties in imagining the failure mode of the individual components; moreover, how the forces transfer through parts to cause the failure. Both the 3D printed model, as well as figures, are necessary to convey the concept of the failure mode to the student. An open-source website [14] was selected to get and to print the parts from a 3D model of the knuckle joint, as illustrated in Figure 3. The students later asked to participate in the assembling process of the elements and to practice the assembly process that had a positive impact on the students' understanding.

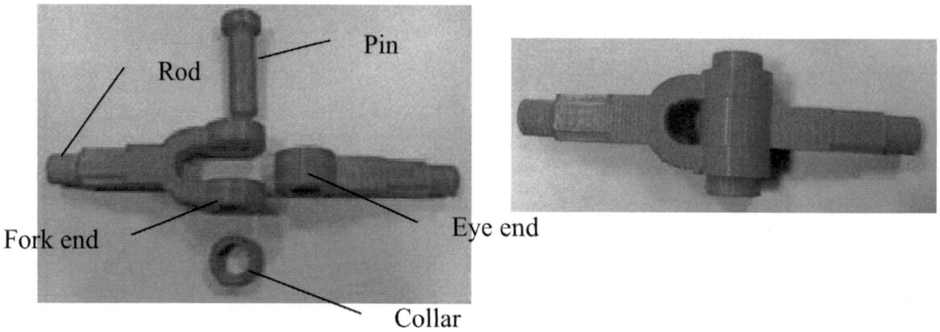

(a) (b)

Figure 3. Disassembled knuckle joint (a), Assembled 3D printed knuckle joint (b).

3. Innovation of Adopting 3D Printing Applications in Machine Design Course

A new assembly training approach based on 3D printing technology was proposed since the assembly is an essential aspect of the manufacturing process. An efficient operation would be achieved through suitable assembly training. Al-Ahmari *et al* [15] proposed an approach to compare existing assembly training methods, including conventional drawing and virtual reality. A 3D printing approach was adopted of a product to evaluate and validate the suggested product. The training performance is assessed for 25 participants using the three-assembly training plan, and it was estimated based on the completion time of assembly task, number of assembly errors, number of frustration points during the job, and completion percentage. It was observed that the 3D printed model performed better than the other two conventional methods. The results also illustrate that there is no significant effect from the 3D model scale variation on the assembly training performance. Teaching mechanical engineering course can be enhanced through using 3D printing capabilities, where students become more excited and enthusiastic when allowed to use 3D printing technologies freely in their course work [16], therefore it is the instructors' responsibility to integrate 3D printing in their course through engaging the students into different activities, starting from research and ending with dealing with open-source 3D printers [17-19]. For teaching machine design and manufacturing lab Mech440, an innovative approach was adopted to stimulate the students toward using 3D printing technology prior the midterm, where a pre-midterm task was delivered to them to design and to build a fixture that can be used to hold and to measure the dimensions a manufactured part. The main objective of the exercise is to consistent repeatability of the measurement since it is essential to calculate the process capability of a traditional milling process as well as CNC milling machine, where the students will collect the measured dimensions to estimate the process capability of the manufacturing process statistically. The evaluation criteria were simplicity, accuracy, measurements repeatability. Students were given two weeks to complete the task. The first stage was to use brainstorming to address innovative solutions that would match the task requirements. Then a detailed explanation should be presented for the prototype that has been done by using CAD software. This part was designed as a holder, it's the base of the product which aim to hold the Vernier caliper into the two small triangles with depth of 10 mm, and the desired piece the can be fit in three inserts in the square shape with 10 mm depth also so that the part can be measured as efficiently as possible, as per instructions, The holders were designed in Y-shape with two different sizes, that should match with the base holder in the same place of the Vernier caliper by keeping a good clearance to maintain accuracy. However, both have the same function to perform the measurement, where the most significant size holder is used to measure the highest point, whereas the smallest holder for measuring the lower parts. The inserts in Figure 4 are designed in three shapes with depth 5 mm to fit the workpiece in all directions so it can be measured to get the right dimensions of the piece, where the insert is going to mount by the base holder part.

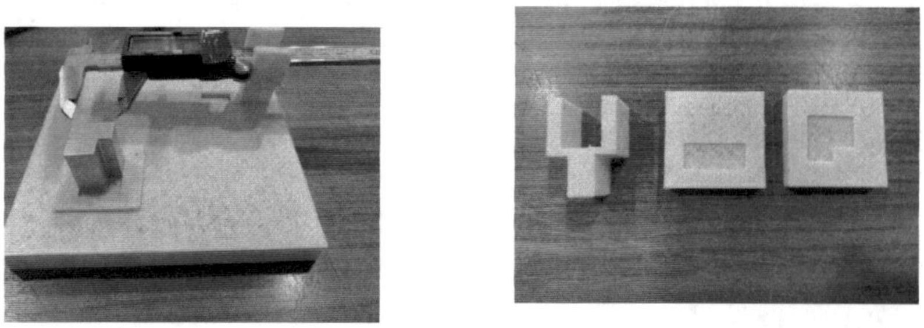

Figure 4. 3D printed holder with the inserts that used for measuring the dimensions.

4. Conclusion

The revolution of the 3D Printing technology has been implemented in the educational sector since the technology has been launched in the market. Although adopting 3D printing technology is fast as the development in the industrial area due to many reasons, but significant efforts have been made to implement the technology through the learning process for different levels. One of the educational applications where 3D printing technology has been implemented in mechanical engineering courses to highlight the contribution to integrating 3D printing technology for the courses in the mechanical engineering department of the UAE University covering the main experimental procedures where they integrated and implemented the 3D printing technology in the educational process. Eventually, by incorporating the 3D printing in the educational process would enhance the educational process including the intellectual properties [20-21] through building a functional 3D printed prototypes that would be tested and proved before manufacturing the full-scale system.

5. References

[1] Ahmed W and Yarub Al-Douri 2020 3D Printing of Ceramic Powder Technology *Metal Oxide Powder Technologies: Fundamentals, Processing Methods, and Applications* (ELSEVIER) Chapter 17.

[2] Ahmed W and Al-Douri Y 2018 Additive Manufacturing Technology for Nanoscale Applications: The Revolution of 3D and 4D Printing Technology in Nanoscale Applications *Meeting on Nanotechnology: Principles and Applications* University of Malaya (UM) KL Malaysia.

[3] Aldarmaki A *et al* 2019 Designing and Developing Innovative Structural Engineering Failure Experiment Using Additive Manufacturing Technology *Advances in Science and Engineering Technology International Conferences* 1-6, doi.org/10.1109/icaset.2019.8714208

[4] Al-Shamsi A *et al* 2019 Assessing Commercial 3D Scanners to Reproduce Structural Elements *Advances In Science and Engineering Technology International Conferences* Dubai, United Arab Emirates.

[5] Awad A *et al* 2019 Designing an Innovative CubeSat Payload to Investigate Material Properties for UAE Space Missions *Advances in Science and Engineering Technology International Conferences, Dubai, United Arab Emirates* doi: 10.1109/ICASET.2019.8714273

[6] Ahmed W and Alhamad I 2018 3D printing innovations in UAE: Case study: Abu Dhabi summer challenge 2017 *Advances in Science and Engineering Technology International Conferences, Abu Dhabi* 2018 doi: 10.1109/ICASET.2018.8376924

[7] Alhamad I *et al* 2019 Boosting Teaching Experience in Mechanical Engineering Courses Using Additive Manufacturing Technologies *Advances in Science and Engineering Technology International Conferences (ASET)* Dubai, United Arab Emirates.

[8] Coakley M and Hurt D 2016 3D Printing in the Laboratory *Journal of Laboratory Automation*, 21(4) 489-495. doi: 10.1177/2211068216649578

[9] Philpot T 2013 Mechanics of Materials: An Integrated Learning System (3rd e.). Wiley.

[10] Girroir J 2018 3d-printing file formats must evolve with the industry *Machine Design.*

[11] Ahmed W 2019. Retrieved from https://eng.uaeu.ac.ae/en/departments/eru/profile.shtml?email=w.ahmed@uaeu.ac.ae

[12] Gere J and Goodno B 2009 Mechanics of materials (7th ed) Cengage Learning.

[13] Khurmi R and Gupta J 2005 A textbook of machine design. New Delhi: Eurasia Publishing House.

[14] Open Source STL files 2020 Retrieved from https://grabcad.com/library/knuckle-joint--1

[15] Al-Ahmari A *et al* 2018 Evaluation of 3D printing approach for manual assembly training. *International Journal of Industrial Ergonomics* 66, 57-62. doi:10.1016/j.ergon.2018.02.004

[16] Alhamad *et al* 2019 3D Printing Applications in Mechanical Engineering Education *3D Printing Applications in Mechanical Engineering Education.* Leiden, The Netherlands: Brill | Sense. doi: https://doi.org/10.1163/9789004415133_006.

[17] Mansour A *et al* 2020 Evaluating the 3D Printing Capabilities Oral Case Studies *International Conference on Industrial Engineering and Operations Management (IEOM)* Dubai, 2020.

[18] Ahmed et al Open Source 3D Printer: A Case Study *International Conference on Industrial Engineering and Operations Management (IEOM)* Dubai UAE.

[19] Al Khawaja H *et al* 2020 Investigating the Mechanical Properties of 3D Printed components, *Advances in Science and Engineering Technology International Conferences (ASET)* Dubai, United Arab Emirates.

[20] Ahmed W and Aziz M 2020 Educational system and method for teaching mechanical failure *US Patent and Trade Marks Office* US10580322B1.

[21] Ahmed W 2019 Toothpaste cap with dental care tools *US Patent and Trade Marks Office* US10399747B2.

IOP Conf. Series: Materials Science and Engineering **938** (2020) 012003 doi:10.1088/1757-899X/938/1/012003

The Application of Harmonic Analysis Methods to Modeling and Simulating the Deformation of the Ring Shell of a Rotating Furnace under High Temperature Influence

Pankratov V.M., Golikov A.V., Barulina M.A., Pankratova E.V., Markelova O.V.

Institute of Precision Mechanics and Control, Russian Academy of Sciences, Saratov, Russian Federation, 410028, Saratov, 24 Rabochaya str., Russia

marina@barulina.ru

Abstract. In the paper, the problem of calculating the deformed state of the inner ring shell of a rotating furnace under high temperature is solved by using harmonic analysis methods. The constructed mathematical model allows us to obtain parameters of furnace's shell deformation based on an array of displacement, which is measured at the reference points. The reference points define the shape of the shell in the polar coordinate system in a discrete way. The results of the work allow automated processing of the metrological data in order to develop practical recommendations of the constructive and methodological way for the complementation of structural elements and equipment units or thermal control systems, if necessary.

1. Introduction

The leading technical equipment for the production of the various kind of friable materials is a rotating furnace [1-3], which used in the manufacture of mortar, materials and some consumable raw materials that required drying. That type of equipment may have a different design, size, and unique characteristics. The primary purpose of rotating furnace – is to produce cement clinker by burning dry and wet raw mixture(slimes). The rotating furnaces are quite complex industrial equipment, which is both physical and chemical reactor and furnace chamber. During the work process of such a furnace, the deformation of the ring shell occurs regularly. For example, it occurs due to high-temperature influences, which lead to bending of the geometrical axis of the shell and, consequently, to the irregularity of distribution of applied loads. The non-regular distribution of the applied loads is the main source of some dangerous accidents [4-8]. Therefore, the necessity of control of temperature and mechanical processes inside and outside of the furnace is the actual problem. Various methods exist for the process of monitoring of thermal deformation [6,7]. One of these methods is to obtain a discrete array of deformations from laser sensors, measuring the deviation of some points of the thin shell from its nominal value.

In this work, the problem of obtaining the parameters of the deformed state of the inner ring shell of the furnace is solved based on empirical input data and in consideration of the high-temperature effects. For solving the formulated problem, the mathematical model of furnace's deformation was created based on methods of the harmonic analysis.

2. Mathematical model

IOP Conf. Series: Materials Science and Engineering **938** (2020) 012003 doi:10.1088/1757-899X/938/1/012003

The input data for solving this problem are:
- nominal radius r_0 of the thin ring shell,
-the inner radius of shell ring r_1,
- the array of fiducial (reference) points $r_i(\varphi_i) = r_0 + \Delta r_i(\varphi_i)$, $(i = \overline{1, N}$, $N \geq 36)$ describes the shell's shape in the polar coordinate system (fig.1).

Herewith, deviations $\Delta r_i(\varphi_i)$ from the nominal radius are small (for example, up to 10%), thus $\Delta r_i / r_0 << 1$.

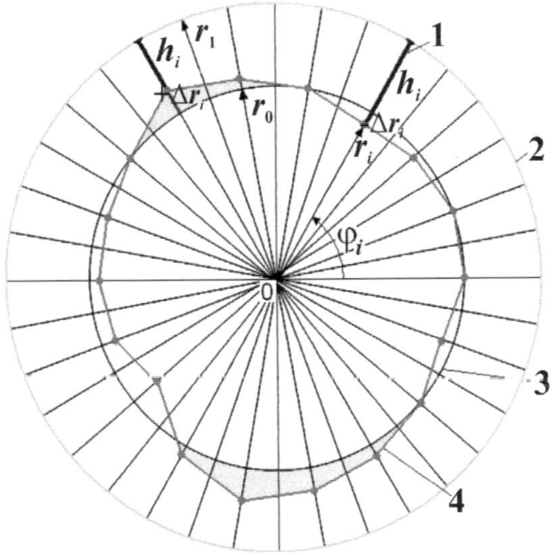

Figure 1. Deformed state of shell:
1 – locking device; 2 – ring shell; 3,4 – non-deformed and deformed shells.

An array of reference points is formed from data received from laser sensors, which make measurements of the changeable value r_i during the furnace rotation. The frequency of the measurements is specified.

The output data is:
- coefficients a_n, c_n, $n = \overline{1, M}$ of Fourier series;
- value of angles φ_i and appropriate arrays $\Delta R_i(\varphi_i)$, $\Delta R_{i*}(\varphi_i)$ of small radial deviations with their signs in dimensional and dimensionless types and an array $h_i(\varphi_i)$ of intervals from ring shell to shell surface, in case of $i = \overline{1, L}$, $L = 10N$ or $L = 100N$;
- radius R_T of the new circle, appropriate for the deformed shell;
- coordinates x_C, y_C of the geometrical center of the new circle with radius R_T.

The main methodology for solving the problem with the initial data defined above is as follows.

The input array $r_i(\varphi_i) = r_0 + \Delta r_i(\varphi_i)$ reduced to a dimensionless form:

$$r_{i*}(\varphi_i) = \frac{r_i(\varphi_i)}{r_0} = 1 + \frac{\Delta r_i(\varphi_i)}{r_0}. \quad (1)$$

Let us present an array (1) as a Fourier series [6,7].

The Fourier series of M terms for the deformed shell with variable radius $R(\varphi)$, has the form:
- in dimensional form

$$R(\varphi) = R_T + \delta \sum_{n=1}^{M} a_n \cos n\varphi + c_n \sin n\varphi\,; \quad (2)$$

- in dimensionless form

$$R_*(\varphi) = \frac{R(\varphi)}{R_T} = 1 + \frac{\delta}{R_T} \sum_{n=1}^{M} a_n \cos n\varphi + c_n \sin n\varphi\,. \quad (3)$$

The value of the new radius R_T, which is the average value of the numbers of the array (1), and the values of Fourier coefficients a_n, c_n, reflecting the shape of the deformed shell, are calculated using the appropriated definite integrals by the method of Fourier harmonic analysis.
The number M terms of Fourier series (2) or (3) is selected from the condition of accomplishment quite matching up of harmonic function (2) or (3) with the discrete response (1) on expectation function and mean square deviation.
In (2), (3) δ – small parameter, which is necessary for analytical research.
To control the calculation of radius R_T of the new circle corresponding to the new deformed shell (1), using the least square adjustment method (LSAM).
Find the circle with radius R_T, which is the best approximation of the discrete function defined by (1) or the Cartesian coordinate system as:
$x_i = r_i \cos\varphi_i, \quad y_i = r_i \sin\varphi_i, \quad 0 \le \varphi_i \le 2\pi.$
According to the least square adjustment method [9,10], the residual values are calculated:
$x_i^2 + y_i^2 - R_T^2 = \varepsilon_i$.
The function U is introduced and minimized:
$$U = \sum_{i=1}^{N} \varepsilon_i^2 = \sum_{i=1}^{N} \left(x_i^2 + y_i^2 - R_T^2\right)^2 \to \min .$$
Calculating $\dfrac{\partial U}{\partial R_T} = 0 \quad \Rightarrow NR_T^2 - \sum_{i=1}^{N}\left(x_i^2 + y_i^2\right) = 0 \quad \Rightarrow NR_T^2 - \sum_{i=1}^{N} r_i^2 = 0.$
The final formula for the new radius R_T, according to the deformed shell, using the least square adjustment method is:
$$R_T = \sqrt{\frac{\sum_{i=1}^{N} r_i^2}{N}}.$$
The following formulas perform the calculation of the array of the small radial deviations with their signs and intervals h_i from the ring shell to the shell surface using the obtained based on (2) or (3):
- in dimensional form
$$\Delta R_i(\varphi_i) = R(\varphi_i) - R_T = \delta \sum_{n=1}^{M} a_n \cos n\varphi_i + c_n \sin n\varphi_i ; \quad (4)$$
- in dimensionless form
$$\Delta R_{i*}(\varphi_i) = \frac{R(\varphi_i)}{r_0} - \frac{R_T}{r_0} = \frac{\delta}{r_0} \sum_{n=1}^{M} a_n \cos n\varphi_i + c_n \sin n\varphi_i . \quad (5)$$
- in dimensional form
$$h_i(\varphi_i) = r_1 - (r_0 \pm \Delta R_i) . \quad (6)$$
Here $i = \overline{1, L};$, где $L > N$ in specified number of times, for example, $L = 10N$ or $L = 100N$.
Such detailed division (up to $L = 100N$) by the angle φ (4)-(6) allows, if necessary, to take into account the tolerances for installing the locking devices at the specified point of the shell.
We perform analytical and numerical findings of the coordinates x_C, y_C of the geometric center of the new circle with radius R_T.
According to the known provisions [10], the determination of the geometric center x_C, y_C of the new circle with radius R_T is realized through the calculation of double integrals in the polar coordinate system by closed area G by the following formulas:
$$x_C = \iint_G r^2 \cos\varphi dr d\varphi \Big/ \iint_G r dr d\varphi , \qquad y_C = \iint_G r^2 \sin\varphi dr d\varphi \Big/ \iint_G r dr d\varphi ; \quad (7)$$

$$G:\begin{cases} 0 \le r(\varphi) \le R(\varphi), \\ 0 \le \varphi \le 2\pi; \end{cases} \tag{8}$$

In correlations (8), describing a closed area G with the representation (2), and using the representation of $R(\varphi)$ in the form:

$$R(\varphi) = R_T + \delta \sum_{n=1}^{M} a_n \cos n\varphi + c_n \sin n\varphi = R_T + \delta f(\varphi)\,. \tag{9}$$

Calculating integrals in (7) taking into account (8), (9) and neglecting small terms of the second and higher order, we get:

$$x_C = \int_0^{2\pi} \cos\varphi d\varphi \int_0^{R(\varphi)} r^2 dr \Big/ \int_0^{2\pi} d\varphi \int_0^{R(\varphi)} r dr = \delta a_1\,, \tag{10}$$

$$y_C = \int_0^{2\pi} \sin\varphi d\varphi \int_0^{R(\varphi)} r^2 dr \Big/ \int_0^{2\pi} d\varphi \int_0^{R(\varphi)} r dr = \delta c_1\,. \tag{11}$$

As we can see, the displacement x_C (on the horizontal axis x) of the geometrical center defined by the Fourier coefficient a_1 at $\cos\varphi$.

The offset y_C (on the vertical axis y) of the geometrical center defined by the Fourier coefficient c_1 at $\sin\varphi$.

Other Fourier coefficients in (9) are not affected by the geometrical center displacement, which is quite consistent with the general provision of theoretical mechanics and mathematics.

On the fig.2 shows an example of calculating the radial diagram of the ring shell deformation based on the input data.

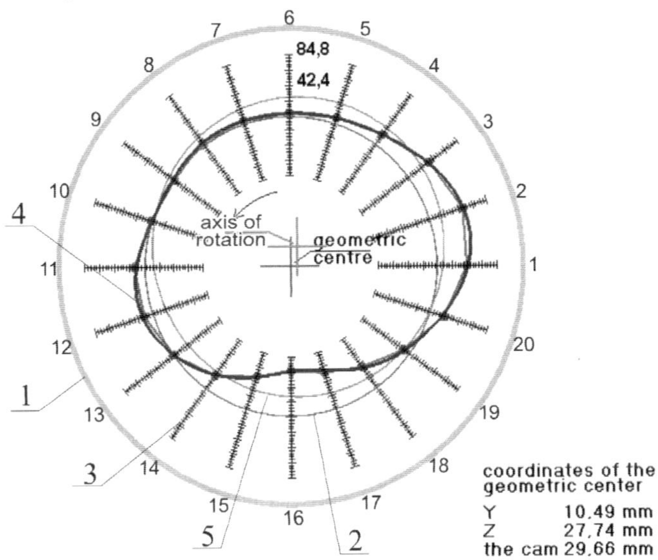

Figure 2. Radial diagram of the furnace shell deformation. 1- shell ring, 2 – non-deformed shell, 3 – deviation scale (mm), 4 – deformation diagram, 5 – the new circle appropriate to the deformed shell

The main calculation results in graphical form are the value of the radial deviation (shell deformation) and displacement of the rotation axis. The numerical data in the table in the figure shows the offset of the geometric center of rotation and the eccentricity.

3. Conclusion.

A mathematical model of the deformed state of the inner ring shell of a rotating furnace is constructed using harmonic analysis methods. The possibility of analytical and numerical findings of the coordinates x_C, y_C of the geometrical center of the new circle with radius R_T was shown. The constructed mathematical model is rather simple and can be implemented both in universal environments for mathematical modeling and in specialized software.

4. Bibliography

[1] Khodorov E I 1968 *The cement industry furnaces* (Leningrad: Stroiisdat) p. 456
[2] Rotary kilns for cement plants Available from: https://flsmidth-prod-cdn.azureedge.net/-/media/brochures/brochures-products/pyro/2000-2017/rotary-kilns-for-cement-plants.pdf [Accessed 1th Marth 2020]
[3] Vaccaro M 2006 Burning alternative fuels in rotary cement kilns *Cement Industry Technical Conference (IEEE Conference Record)* 10
[4] Tuishev A I and Tokarev D G 2015 Methods for determining the deformation state of rotating furnace *Vestnik NSIEI* **12 (55)** 82-84
[5] Sholomitskii A A, Kovalev P S, Medvedskaya T M and Martinov A V 2017 The effect of heating the furnace on the straightness of its axis of rotation *Vestnik SSGA* **22(4)** 18-24
[6] Petrov V V, Turin S V and Kopitov A N 2010 Control of the geometrical characteristics of the rotation furnaces *Cement and its usage* **2** 78–82.
[7] Zhang Y and Zheng Weifei 2011 Dynamic detecting method of the axial distortion and shell distortion of rotary kiln *Cement Engineering* **02** 20-26
[8] Boateng A A 2011 *Rotary Kilns: Transport Phenomena and Transport Processes* (Butterworth-Heinemann)
[9] Bronshtein I N, Semendyayev K A, Musiol G and Mühlig H 2007 *Handbook of Mathematics* (Springer Berlin Heidelberg)
[10] Adler Y P, Markova E V and Granovsky Y V 1976 *Experiment planning when searching for optimal conditions* (Moscow: Nauka)

Acknowledgments

The research was carried out within the state assignment of the Ministry of Science and Higher Education of the Russian Federation (theme No. AAAA-A18-118042790040-0)

IOP Conf. Series: Materials Science and Engineering **938** (2020) 012004 doi:10.1088/1757-899X/938/1/012004

Optimization of Gate Resistance for Motor Driver Design

Y G Cho[1], **J W Baek**[1,2]

[1] Reliability Assessment Laboratory, Korea Construction Equipment Technology Institute, 36 Sandan Street, Gunsan, 54004, South Korea
[2] Corresponding Author

backy@koceti.re.kr

Abstract. Owing to the rapid increase in the use of small- and medium-sized motors at various industry, the reliability of the electronic control systems is emerging as an issue. Accordingly, research on a method that guarantees reliable motion of motors by selecting appropriate gate resistance of the driver part and that contributes as a solution to the thermal issue and EMC(Electromagnetic Compatibility) in terms of design was conducted. Among methods that increase the reliability of motor drivers, one involving the calculation of an optimal gate resistance value is the most efficient in cost aspects without he added hardware or the application of a complex motor control algorithm. Therefore, this study proposed the process of finding optimal gate resistance of the driver circuit. An approach was made to find a safe operation area about switching, which has considered only partially, and a practical experiment was carried out through the evaluation board to prove the proposed method. As a result, although the estimated range and experimental results had slight deviation, we proved that it could be used as a method to find the optimal resistance value.

1. Introduction

Today, the electronic systems of industries have to be designed with goal of reducing expenses while maintaining high reliability. It is very contradiction. In particular, with respect to design considering the reliability of motor controllers, including EMC and hardware composition, many research studies on software composition (algorithm) are in progress[1][2][3][4]. Except software reliability, in hardware reliability, there is a trade-off with the cost. Therefore, this issue should be approached from the aspect of efficiency. It is necessary to find a method that guarantees the stable function of hardware without adding any further hardware components. Although there can be various methods and discussion, in this paper, we have described a method to select an appropriate gate resistance value to optimize the switching properties of a motor driver. This makes it possible to derive a cost-effective benefit of controlling EMC properties and simultaneously guaranteeing reliable operation of the motor[5].

2. Optimization Method of Gate Resistance

The basic structure of bridge circuits can be interpreted in the context of a correlation between the gate driver that input he on/off signals such as PWM to the gate of the switching device and the power device that switches the high voltage or the high current.

2.1. Minimum resistance value for protection of device in the worst case

IOP Conf. Series: Materials Science and Engineering **938** (2020) 012004 doi:10.1088/1757-899X/938/1/012004

In general, a small quantity of current is applied to obtain the Miller capacitance as well as the threshold voltage required to turn on the switching device. As the current is momentary, the switching time(t_{SW}) depends on its quantity. In general, the maximum output for this transient current is specified in the data sheet of the gate drive IC. First, the minimum resistance for device protection in the worst case can be expressed as follows:

$$R_{protect} = \frac{\Delta V_G}{I_{sink}} \tag{1}$$

$R_{protect}$ is the resistance value to avoid the physical damage of the gate drive IC, which is the minimum reference value for all subsequent considerations.

2.2. Resistance value for rapid turn-on

The calculation of the resistance value to facilitate rapid switching when turned on switching device can be examined from two perspectives. Switching when considered from the device viewpoint takes into account he gate input current and the output characteristics of the switching device. The gate input current can be replaced with the gate driver output current. The R_G calculation approach that considers each element is as follows[6]:

$$I_{AV} = \frac{Q_{gd} + Q_{gs}}{t_{SW}} \tag{2}$$

$$\frac{dV_{DS(falling)}}{dt} = \frac{I_{AV}}{C_{rss}} \tag{3}$$

$$R_{G(min)} = \frac{V_{CC} - V_{miller}}{I_{AV}} = \frac{V_{CC} - V_{miller}}{C_{rss} \cdot \frac{dV_{DS(falling)}}{dt}} \tag{4}$$

On the basis of equation (2) and (3), we can derive the following:

$$\left(\frac{Q_{gd} + Q_{gs}}{C_{rss}}\right) \cdot \frac{1}{t_{SW}} = \frac{dV_{DS(falling)}}{dt} \tag{5}$$

In general, as there are many cases where the gate driver lacks the current driving ability, I_{source} is the first input in place of I_{AV} for the sake of calculation. Then, the maximum values of t_{SW} and right term of equation (5) are calculated. However, as right term of equation (5) is a parameter that is already specified in the physical specifications of the switching device, certain selection work is required. If right term of equation (5) in the data sheet has a lower value, I_{AV} and t_{SW} are calculated on the basis of this value and these values can be regarded as the maximum I_{AV} value and the maximum t_{SW} value that can be obtained from a combination of the gate driver IC and the switching device.

2.3. Resistance value for deterministic turn-off

When the switch is opened, the inrush current flows through C_{rss}(Miller capacitance) because of the rapid voltage change[7]. The current path is shown in Figure 1. Further, the gate resistance should be selected by considering the prevention of a phenomenon in which the switch is closed when a voltage higher than the gate threshold voltage is applied because of the current that flows when the switch is opened[6].

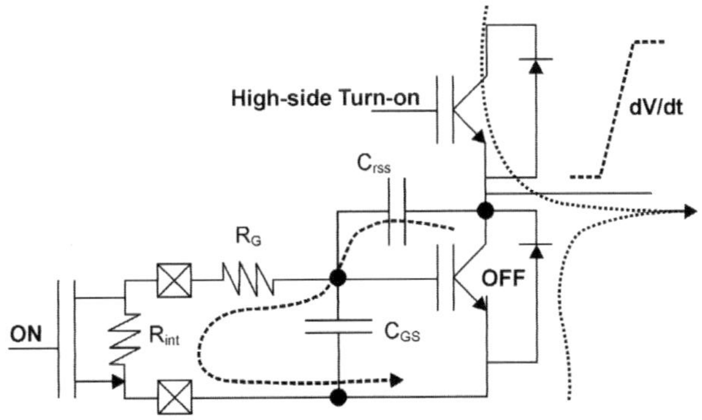

Figure 1. Current path when low-side is off and high-side is on
By summarizing the above contents in the form of a formula, we obtain the following:

$$R_{G(max)} \leq \frac{V_{TH}}{C_{rss} \cdot \frac{dV_{DS(rising)}}{dt}} \tag{6}$$

Here, $R_{G(max)}$ becomes the maximum resistance value for switching the safe operation area(SOA), among the considered values.

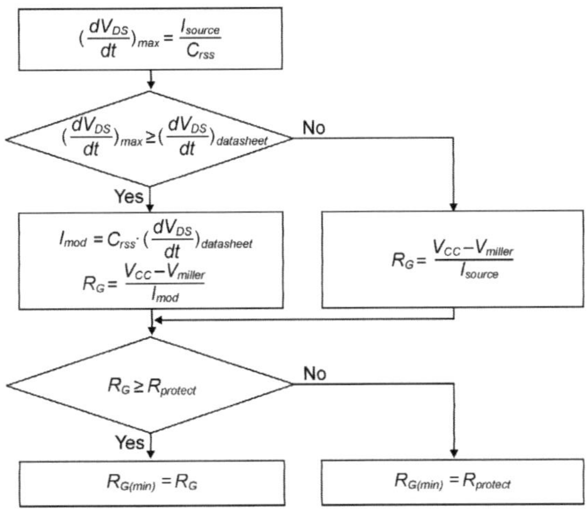

Figure 2. $R_{G(max)}$ calculation flow

Through the considerations in switching procedure of turn-on and turn-off, the flow chart of Figure 2 was derived. This calculation flow simply determines $R_{G(max)}$ by entering parameters of driver IC and switching device.

3. Experimental Results

The experiment was conducted using TI DRV8301-LS31-KIT and controlled by the input of a BLDCM Hall Sensor. A system in which the control is stopped in the case of a hardware error, which is also recognized as a fault in the controlling motor such as arm short through TMS570LS31 supporting ISO26262, was set up.

Figure 3. Preparation for experiment

Furthermore, the current driving capability of DRV8301 is available in three options, and the experiment was conducted after setting up hardware with which rapid switching was possible by setting the largest capability[8].

As the selection of the gate resistor procedure according to the flow chart of Figure 2, first,

$$\left(\frac{dV_{DS}}{dt}\right)_{max} = \frac{I_{source}}{C_{rss}} = \frac{1.7 \cdot 10^{12}}{600} \approx 2.833\ V/ns \tag{7}$$

The MOSFET of DRV8301-LS31-KIT is SUM110N06-3m9H, whose $t_{f,max}$ is 14 ns @ 30 V, and after adjusting units, $t_{f,max}$ is about 2.143 V/ns[8][9]. The maximum output voltage slope of the switching device is more gradual, and the combined properties of the gate driver and the MOSFET can be regarded as considerably better than that of the MOSFET. R_G can be found by calculating the current by using the value of 2.143 V/ns.

$$I_{mod} = C_{rss} \cdot \left(\frac{dV_{DS}}{dt}\right) = 600 \cdot 10^{-12} \cdot 2.143 \cdot 10^{9} \approx 1.286 \tag{8}$$

$$R_G = \frac{V_{cc(max)} - V_{miller}}{I_{source}} = \frac{11.5 - 6.3}{1.286} \approx 4.044 \tag{9}$$

Another method that can be considered for obtaining the minimum gate resistance is the calculation of the resistance value required for protecting the gate driver in the worst case.

$$R_{protect} = \frac{\Delta V_G}{I_{sink}} = \frac{11.5}{2.3} = 5 \tag{10}$$

$$R_{G(max)} \leq \frac{V_{TH(min)}}{C_{rss} \cdot \frac{dV_{DS(rising)}}{dt}} = \frac{3.4}{600 \cdot 10^{-12} \cdot 187.5 \cdot 10^{9}} \approx 30.222 \tag{11}$$

Here, we conclude that R_G should have a value in the range 5 to 30.222 Ω.

Figure 4 shows the result of the measuring waveforms in accordance with the resistance values obtained through an experiment. The waveform seems to change stably with an increase in the resistance value; however, at values higher than 17 Ω, frequent switching errors cause the controller to sense faults.

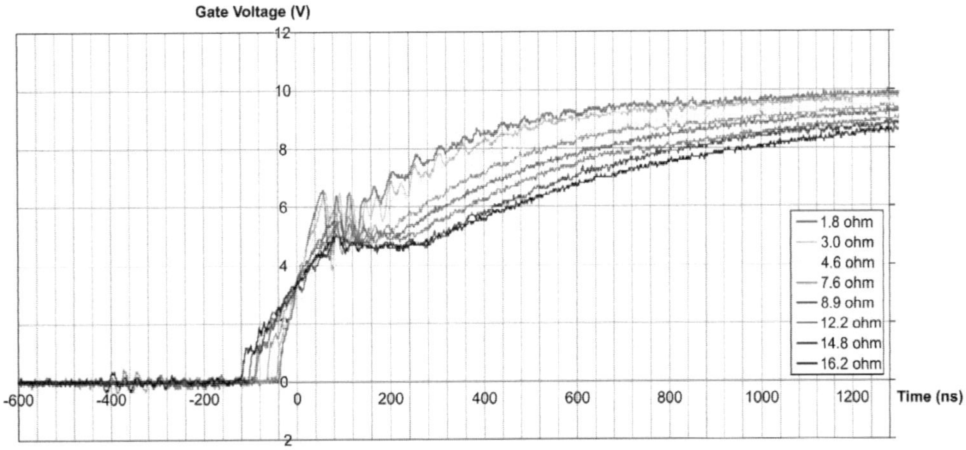

Figure 4. Waveforms of switching device input

As Figure 4 shows, it may be recommended to select a gate resistance from the SOA range of 7.6 to 16.2, and it would be better to select smaller value within the range because devices are being studied to enable high-speed switching performance these days[10]. The experimental results show that 8 to 9 ohms is thc optimal setting value because it is advantageous for the gate resistance value to be lowered in order to maintain on-state of the MOSFET[11].

4. Discussions and Future Work

Thus far, we have discussed the calculation of the gate resistance value required for improving the reliability of the motor driver part, and in safety-required electronic control system, every device has to be selected carefully. In particular, the electronic components controlling a motor can enter into the failure state easily because of switching and a rush current. Further, the failure state can fatally decrease the reliability of the entire system. It is anticipated that a more appropriate gate resistance can be calculated through the proposed method. The theoretical calculations based on the parameters of the devices available through the datasheet may not exactly match the practical experiments. The reason is that in addition to the two dominant devices, the characteristics of PCB patterns and the source or ground characteristics and conditions are very different in many cases. However, trying to estimate SOA through the proposed method is a meaningful approach and the results of the experiment are proving it. Further, the findings of this study are expected to contribute to EMC solution and heat property improvement[12]. Further, the findings of this study are expected to contribute to EMC solution and heat property improvement. I expect that other significant methods can be found, analyzing the correlation between the amount of EMI variation and switching loss due to changes in resistance values.

5. References

[1] V Bolloju and J Yang 2011 Influence of short circuit conditions on IGBT short circuit current in motor drives *26th Annual IEEE Applied Power Electronics Conference and Exposition* 10.1109/APEC.2011.5744820

[2] J Tyster, M Iskanius, J P Strom, J Korhonen, K Rauma, H Saren and P Silventoinen 2009 Hight-speed gate drive scheme for three-phase inverter with twenty nanosecond minimum gate drive pulse *13th European Conference on Power Electronics and Applications* (AN)10939658

[3] H Sawezyn, N Idir and R Bausiere 2002 Lowering the drawbacks of slowing down di/dt and dv/dt of insulated gate transistors *2002 International Conference on Power Electronics, Machines and Drives* 10.1049/cp:20020176

[4] P J Grbovic 2005 The gate drive requirements for new generation of high current low voltage IGBTs employed in motor converters *2005 International Conference on Power Electronics and Drives Systems* 10.1109/PEDS.2005.1619773

[5] R J Valentine 1996 IGBT High frequency operation *1996 IEEE Industry Applications Conference 31th Annual Meeting* 10.1109/IAS.1996.559259

[6] A Merello, A Rugginenti and M Grasso 2004 Using monolithic high voltage gate drivers *International Rectifier*

[7] S Azzopardi, A Benmansour, J M Vinassa, O Briat and E Woirgard 2005 Effect of the high input capacitance of 1200V trench IGBT on the switching characteristics under inductive load *2005 European Conference on Power Electronics and Applications* 10.1109/EPE.2005.219278

[8] Texas Instruments 2011 DRV8301 Datasheet

[9] Vishay Siliconix 2018 SUM110N06-3m9H Datasheet

[10] T Nomura, H Kambayashi, M Masuda, S Ishii, N Ikeda, J Lee and S Yoshida 2006 High-temperature operation of AlGaN/GaN HFET with a low on-state resistance, high breakdown valtage, and fast switching *IEEE Transactions on Electron Device* vol 53, pp 2910–2911

[11] M Berkani, S Lefebvre and Z Khatir 2013 Saturation current and on-resistance correlation during repetitive short-circuit conditions on SiC JFET transistors *IEEE Transactions on Power Electronics* vol 28, p 622

[12] M Inaba, J Sakano, H Miyazaki, M Iwamura and Y Maeda 2003 A high reliability and low loss 3-phase gate driver IC with a novel soft gate drive circuit for 42V motor generator system *2003 IEEE 15th International Symposium on Power Semiconductor Devices and ICs*

IOP Conf. Series: Materials Science and Engineering **938** (2020) 012005 doi:10.1088/1757-899X/938/1/012005

Electroencephalogram (EEG) Signal Classification Using Artificial Neural Network to Control Electric Artificial Hand Movement

A S Saragih[1]*, A Pamungkas[1], B Y Zain[1], W Ahmed[2]

[1]Department of Mechanical Engineering, Faculty of Engineering, Universitas Indonesia, Depok, 16424, Indonesia.
[2]Engineering Requirements Unit, College of Engineering, Al Ain, Abu Dhabi, United Arab Emirates University

*Corresponding author's : ashamsuddin@ui.ac.id

Abstract. All due to the complex nature of the electroencephalography (EEG) signal, it is a challenge to be able to use it as the driver of an electric artificial hand. By using EEG signal, the command for artificial hand movements becomes more intuitive and natural. This study aims to classify EEG signals to serve as electronic hand control. Classification is conducted using artificial neural networks (ANN), in which EEG signal datasets are obtained from a commercial brain computer interface (BCI). The ANN model obtained is expected to be able to determine that the EEG signal is one of the five EEG signals generated from five predetermined hand movements. This study proposes feature extraction and processing that is very simple but performs well, indicated by its small error value. The results show that ANN can classify five hand movements tested with an overall accuracy rate of 80%.

1. Introduction

Prosthetics are artificial devices that replace lost body parts, which are caused by trauma, disease or congenital conditions [1]. Prosthetic hands or artificial hands are grouped into two main groups, namely cosmetic and functional artificial hands. Cosmetic hands are made to resemble the looks of the original arm without any basic hand functions. Functional artificial hands are artificial arms that can perform several arm functions, ranging from simple functions to complex functions [2].

The development of artificial hand technology lies in the fields of actuators, sensors, controllers and fabrication. Hand control systems that have been developed include electromyography (EMG) and the EEG. Control systems using EMG contractions of the muscles in the arms produce raw EMG signals, in which raw EMG signals are manifestations of neuromuscular activity associated with muscle contraction [2]. Artificial hand control using EMG signals has limitations in some cases such as inconsistency problems, less intuitiveness or must be done in a patient state and requires muscle work. Research on the use of EEG signals as artificial hand controls uses facial expressions [3, 4]. Certain facial expressions such as blinking will generate an EEG signal that will be recorded by the sensor. Then this signal is processed to move the artificial hand.

This control system has weaknesses, including facial expressions having limitations which therefore limits the artificial hand control. The EEG signal produced by facial expression, in principle, is not an EEG signal to move an artificial hand. Thus, the control system is less natural. The EEG signal from facial expressions that has been processed by the system in the headset EEG becomes a mature signal. To get control of natural limbs originating from the brain, user factors and design preferences are a priority in the process of translating brainwaves into commands in moving artificial hands. Therefore, the purpose of this study was to classify electroencephalogram (EEG) signals originating from hand movements using artificial neural networks. After the classification is successfully carried out, the network can determine the type of hand movements that are tested randomly. The output of this network is expected to be used as an input command to move artificial hands, so that electronic artificial hand control can be achieved using EEG signals. By using EEG signals, the command for artificial hand movements becomes more intuitive and more natural.

2. Materials and methods

The research method consists of two stages, training and testing. At the training stage, the subject performs hand movements in accordance with medical rehabilitation procedures. Five commonly used hand movements; cylindrical grip (CG), power handgrip (PG), tripod pinch (TP), index point (IP), and resting (open grip-OG) are performed. A movement produces an EEG signal, which is then captured by EEG sensor OpenBCI^R Cyton to obtain raw signals. This raw signal is processed through feature extraction stages, in which the obtained EEG signal is divided into training data and test data. Training data is used as input in the classification training process. The method used for the classification process is artificial neural network (ANN). The classification class target is determined based on the type of hand movement according to Table 1. The output parameters obtained by the network from the EEG signal classification process are used for the testing process.

Table 1. Type of hand movement.

Item	CG	PG	TP	IP	OG
train data	1-9	10-18	19-27	28-36	37-45
target	0.00	0.25	0.50	0.75	1.00

The design of the ANN architecture begins with de-signalization process, then changes are made to the training parameters such as the number of hidden layers, the number of epochs / iterations and the rate of learning. Evaluation of classification results is done by determining the accuracy value according to the following equation [5]:

$$accuracy = \sum_{i=1}^{l} \frac{tp_i + tn_i}{tp_i + fn_i + fp_i + tn_i} \tag{1}$$

Where tp_i, tn_i, fn_i, and fp_i are obtained from the confusion matrix shown at table 2 as follows:

Table 2 Confusion matrix for binary classification.

Data class	classified as positive	classified as negative
Positive	true positive (tp)	false negative (fn)
negative	false positive (fp)	true negative(tn)

The ANN used for the classification of EEG signals is multi-layer neural network backpropagation. The input pattern used in the network is a dataset of five movements carried out by the test subjects as shown in Figure 1. The network is designed using one input layer consisting of 48 neurons obtained from the extraction of six features and each represents the 8 channels of EEG sensor OpenBCI^R Cyton. There are three hidden layers, with the number of neurons determined by trying several configurations until high performance or a low error value is achieved. One output layer consists of one neuron. The activation function used is a binary sigmoid function.

The next step is to determine the network training function. Perfect performance is achieved if the error value is zero. Since that is difficult to achieve, the error value close to zero can be categorized as having a good performance. Experiments have been conducted using nine training functions from Matlab^R software.

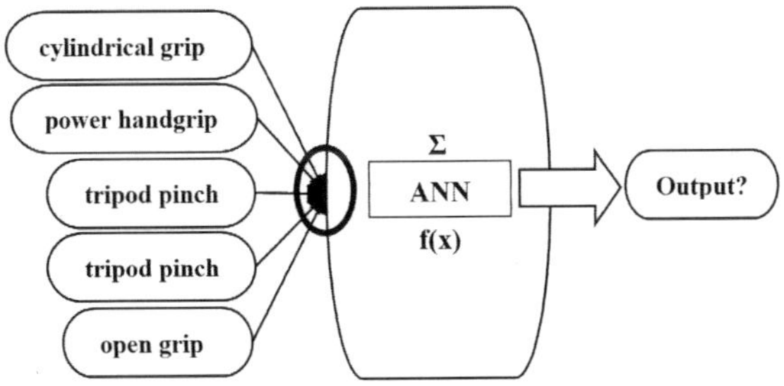

Figure 1. Illustration of artificial neural networks for classification of hand movements.

The previous experimental results showed that the optimal training function to be used in the classification process is the function of trainlm. Trainlm is a Levenberg-Marquardt (LM) algorithm that is classified as a high-order adaptive algorithm that can minimize Mean Square Error from a network. The function 'trainlm' is also widely used in several studies and has proven effective as a learning function. By using the training function and the same training parameters, the EEG signal classification process is carried out on 10 subjects.

3. Results and discussion

The others training functions of traincgb, traincgf, traincgp and trainrp have low performance that during the training process, the training function gradient value or minimum step has been reached before the performance goal is reached.

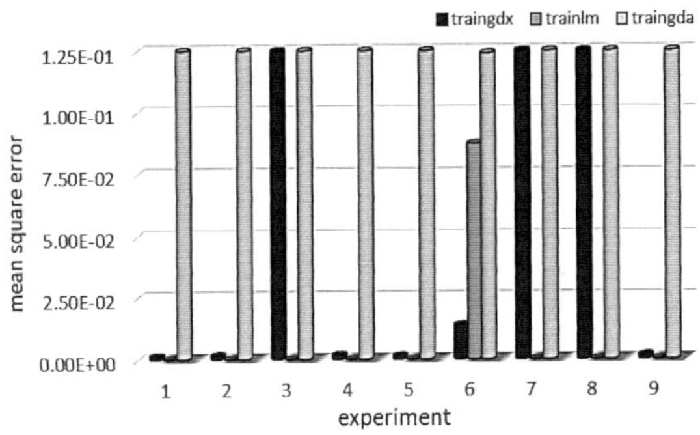

Figure 2. Comparison of the error values of traingda, traingdx and trainlm.

Figure 2 shows that the training function traingda produces similarly low performance even though the maximum epoch value has been reached. Traingdx and trainlm managed to get a pretty optimal performance. Based on the results of this experiment the 'trainlm' will be used on the network for the process EEG signal classification for 10 subjects.

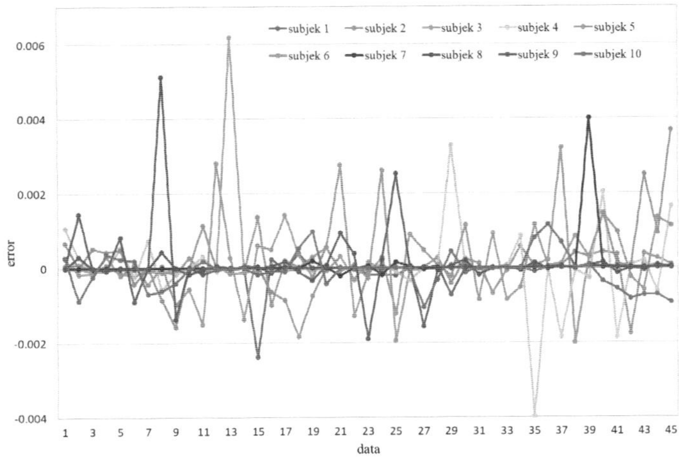

Figure 3. Graph of network training error value.

By using the training function and the same parameters, the EEG signal classification process is carried out on 10 subjects. Figure 3 shows that out of 10 subjects who were trained, the network training error value was almost close to zero. There is a significant deviation in some data as happened in subject 3 and subject 9, but still within tolerable limits. The 13th data of subject 3 has an error value of 0.00618, while the 8th data of subject 9 has an error value of 0.00512. The cause of the error is because the data has a small suitability with the target data that has been set. In the picture it can also be seen that subject 1 and subject 8 have an error value close to zero which is relatively stable.

3.1. Network testing

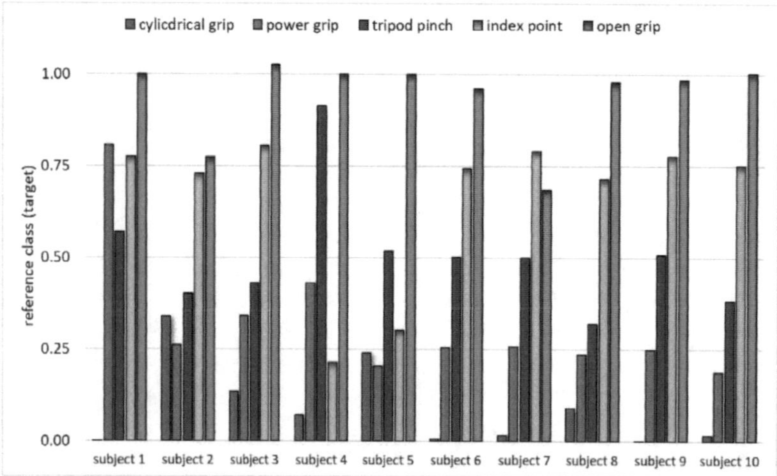

Figure 4. Test Simulation results using test data.

The results of network testing using test data are shown in Figure 4. Overall, the simulation results on the test data are quite successful. For example, in the first test subject, the first movement test data was identified by the network according to the movement class ‘cylindrical grip’. The second movement test data that should have been recognized as a ‘power grip’ movement by the network, is recognized as the ‘index point’ movement. The third movement test data is also able to be correctly recognized as a ‘tripod pinch’ movement, as well as the fourth movement test data that can be correctly recognized as an ‘index point’ movement, even though there is a difference in the output value compared to the target value. The fifth movement test result, the ‘open grip’ movement, is very well recognized with an output value that is very close to the target value.

3.2. Artificial hand testing

The design of artificial hand is developed by improving the anthropomorphically. The actuation of the artificial hand fingers is achieved by tendon mechanism. The servos are controlled by microcontroller with custom proto shield located near the back of the hand.

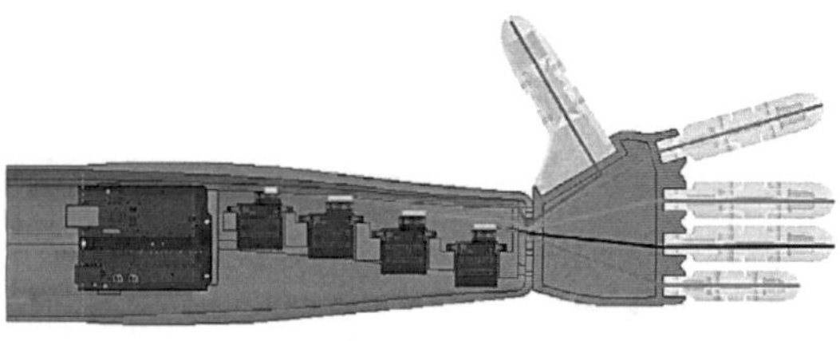

Figure 5. Tendon mechanism of hand design.

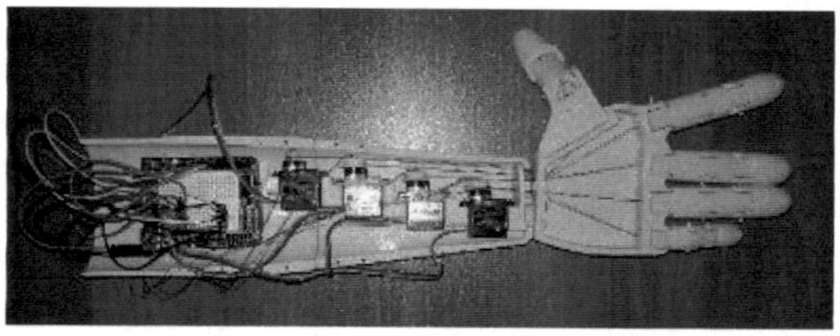

Figure 6. Assembled artificial hand.

The dimension of the hand, with all fingers open, is 143 mm long, 120 mm wide, and 15 mm thick. For the forearm, it is 274 mm long, 79 mm wide, 49 mm thick, and 76 mm thick at battery slot. The weight of the prosthetic hand with servo and microcontroller is 463 grams, while the total weight including batteries is 625 grams. The prosthetic hand is manufactured through FDM 3D printing using PLA+ material. Here, 3D printing manufacturing method is given a special emphasis, especially in relation to production of prostheses. Human hands can greatly differ from one person to another, according to gender, age, race, and any other factors. In relation to the production of prosthetic hands and prostheses devices in general, 3D printing increases versatility of the production process of prostheses devices, especially even though they are complex and highly customized objects, designing and production time can be greatly shaved.

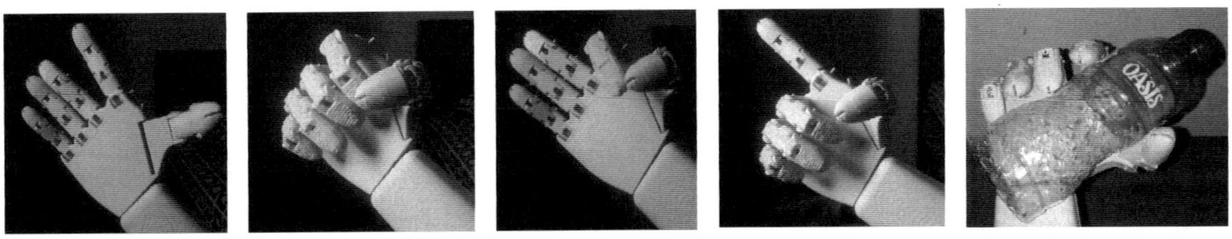

Figure 7. Attempt to perform open grip (top left), power grip (top middle), tripod pinch (top right), index pointing (bottom left), cylindrical grip (bottom right) by artificial hand.

After all the hand components, tendon, servo motor, cables and microcontroller have been installed, the created artificial hand is ready for experimental functionality testing. The tendon used for the artificial hand is a 0.14 mm diameter-fishing line. Fishing line is used as the tendon since it does not stretch when it is pulled, rather, it keeps its tension. In comparison to the predetermined grip postures, the created artificial hand can perform all of the 5 postures as shown in Figure 7.

4. Conclusions
Classification of electroencephalography (EEG) signals generated from hand movements was carried out using artificial neural networks (ANN). The test results show that the ANN model obtained is able to determine the type of hand movement tested from randomly determined test data. The classification of five hand movements tested has a total accuracy level of 80%. Thus, it shows that the classification of EEG signal hand movements can be done through a simple process without needing long signal processing procedure. The Classification was using backpropagation artificial neural networks with trainlm as the training function. The feature extraction used are minimum value, maximum value, mean value, median value, mode value and standard deviation value. The result lay on a specific electrodes configuration of the EEG sensor. Change of the configuration may result different accuracy.

5. References
[1] M. C. Mohan and M. Purushothaman, "Design and fabrication of prosthetic human hand using eeg and force sensor with arduino micro controller," in 2017 Third International Conference on Science Technology Engineering & Management (ICONSTEM), 2017, pp. 1083-1086.

[2] O. B. Lőrinczi and P. Aradi, "Development of a prosthetic hand regarding complex motion and controllability" J Periodica Polytechnica Mechanical Engineering, vol. 55, no. 2, pp. 101-104, 2011.

[3] D. Bright, A. Nair, D. Salvekar, and S. Bhisikar, "EEG-based brain controlled prosthetic arm," in 2016 Conference on Advances in Signal Processing (CASP), 2016, pp. 479-483: IEEE.

[4] M. A. A. Kasim et al., "User-friendly labview gui for prosthetic hand control using emotiv eeg headset," Procedia Computer Science, vol. 105, pp. 276-281, 2017.

[5] M. Sokolova, G. Lapalme, and Management, "A systematic analysis of performance measures for classification tasks," Information Processing and Management, vol. 45, no. 4, pp. 427-437, 2009.

Acknowledgments
This initial research works was funding by Direktorat Riset dan Pengabdian Masyarakat (DRPM) Universitas Indonesia under PITTA-A program no.NKB-0453/UN2.R3.1/HKP.05.00/2019.

IOP Conf. Series: Materials Science and Engineering **938** (2020) 012006 doi:10.1088/1757-899X/938/1/012006

Eccentric Interaction between Strong and Weak Axes in the Design of Asymmetric Angle Steel Strength

W T Hsu [1], S T Chung [2] and R Y Liu [3]

[1] Associate Professor, Department of Construction Engineering at Chaoyang University of Technology, Taichung, Taiwan.
[2] Postgraduate student, Department of Construction Engineering at Chaoyang University of Technology, Taichung, Taiwan.
[3] Civil engineer, Jian' an Civil Structure Technician, Taichung, Taiwan.
wthsu@cyut.edu.tw

Abstract. Steel structural members use many types of material sections and a wide range of sizes. Among them, Angle steel is more economical than I-type due to the use of materials, and is usually used for beam web support and diagonal support. Therefore, this study will analyze the eccentric axis strength change of L-shaped angle steel under the action of biaxial bending moment. The angle steel is divided into equilateral angle steel and unequal angle steel, and the calculation of unequal angle steel is the most complicated. Due to the asymmetry of the axial force, when the angle steel bears an axial force, it will cause eccentric load, which makes it the bending moment caused by the angle steel will change the structure. Therefore, in this study, an unequal-sided angle steel with a length of 10 ft. L8 × 4 × 7/16 was selected as the research object, and the eccentric strength changes of the strong and weak axial forces of the angle steel were analyzed.

1. Introduction

This study is to analyze the eccentric strength of the bending moment on the angle steel when the angle steel is connected to the panel. When the distance between the strong axis (W_D) and the weak axis (Z_D) changes (refer with Figure1), the pressure intensity of the eccentric axial that the angle steel can withstand will also change. To analysis the strong and weak axis, the properties of the angle steel need to be obtained first, and then the correct eccentric pressure intensity is calculated using the AISC2017 standard formula, and then the calculated values are compared and analyzed. Because of the asymmetry of the wheelbase of the unequal angle steel and its calculation is more complicated than that of the equilateral angle steel, this study takes the unequal angle steel L8 × 4 × 7/16 as the research object. The eccentric load design of single angle steel mostly adopts the code formula developed by the American Society of Steel Structures [1], and then studies and analyzes. Because of the reliability of the code and the relevant specifications of the current steel structure design in Taiwan are based on the American Steel Structure Society Therefore, this study uses the AISC (2017) specification as the research basis for analysis of subsequent studies.

Liu [2] tested the behavior of single-angle steel beams and columns using unequal-angle angle steels to test eccentric compression with respect to any major or minor principal axis of the angle steel section. **Madugula** [3,4] studied the design of angle steel and the flexural strength of columns. **Bhilawe** [5] uses equal angles and a range of maximum principal slenderness ratios to determine elastic and inelastic failure characteristics, including flexural torsional effects. **Yuan** [6] the basic assumption used in the research is that the angle of the total strain energy is that part of the beam undergoes a uniformly distributed load can be simplified into a two-stage process. In order to verify

IOP Conf. Series: Materials Science and Engineering **938** (2020) 012006 doi:10.1088/1757-899X/938/1/012006

the analytical solution developed, non- Linear finite element analysis. **Chan** [7] used a simplified formula to calculate the elastic deflection and flexural buckling load of a single-angled member. For single-angled members, a new buckling curve considering the Eurocode 3 design rules was proposed to surround the main and counter-spindles. Withstand eccentric loads. A reliable and effective finite element model is provided to verify the proposed equations and study the different parameters affecting buckling behavior. **Kim** [8] studied the cross-sections of various composite angles to maximize the stress intensity of compression bending under high strength.

Bradford [9] are all studying the bending, torsion and buckling behavior of angle steel members, and studying the strength values they bear. The former two use cold-pressed members. Research experiments were conducted, and the last one was studied with high-strength steel plate beams.

Shi [10] Experimental studies were performed using the sectioning method. The residual stress magnitude and distribution of 15 sections were obtained, and the influence of the aspect ratio was clarified. Further research on the residual stress and buckling behavior of high-strength steel members provided useful experimental data and calculation methods. **Liu** [11] studied the local buckling behavior of high-strength isometric steel under axial compression, using finite element model to analyze the local buckling behavior of high-strength steel isometric steel under axial compression, and verified by test results.

The analysis of single-angle steel has many conservative and underestimated behaviors. This study analyzes the biaxial eccentric pressure strength analysis of single-angle steel, collects its design formula, and explains the analysis steps.

Since the isometric steel differs from the unequal angle steel in that the isometric steel has a symmetry axis and the unequal angle steel has no symmetry axis, the section characteristics of the unequal angle steel are calculated under equal pressure. Complexity. The eccentric load diagram of the angle steel is shown in Figure 1. The D point is the position of the eccentric load, and the ABC three points are the control points of the design strength. According to the recommendations of the AISC specification, the relative position of the load geometry is shown in Figure 1, and analyzed with reference to specific geometries.

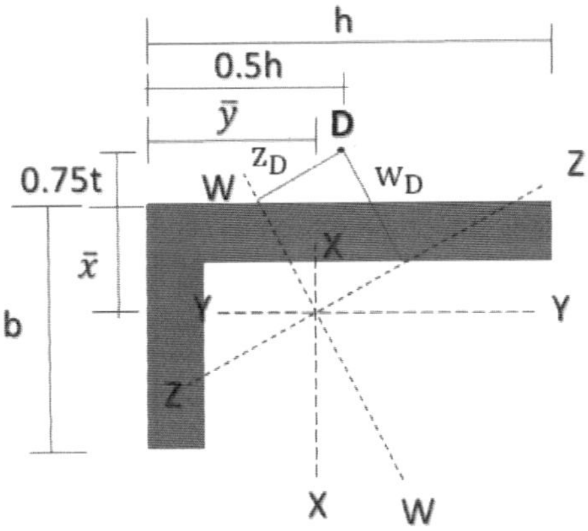

Figure 1. Angle steel point diagram.

2. Eccentric Load Acting on Single Angle Steel (AISC 2017)

Mutual Force Bending Moment Mutual Equation of Integral Member. The difference between the integral component equation and the section equation is that the value of $\frac{P_r}{\phi_C P_n} < 0.2$ must be met after the value is obtained. If the value meets the specification, the component is a large axial force criterion, and otherwise Small axial force criterion.

When $\frac{P_r}{P_c} \geq 0.2$ is the large axial force criterion.

$$\frac{P_r}{P_c} + \frac{8}{9}\left(\frac{M_{rx}}{M_{cx}} + \frac{M_{ry}}{M_{cy}}\right) \leq 1.0 \tag{1}$$

When $\frac{P_r}{P_c} < 0.2$is a small axial force criterion.

$$\frac{P_r}{2P_c} + \left(\frac{M_{rx}}{M_{cx}} + \frac{M_{ry}}{M_{cy}}\right) \leq 1.0 \tag{2}$$

formula：

P_r =Considering the axial compressive stress value of LRFD load combination，kips(N)
$P_c = \phi_c P_n$ =Design axial strength in beam and column，kips(N)
M_r =Consider the bending stress value of the LRFD load combination，kip − in(N − mm)
$M_c = \phi_b M_n$ =Design bending moment strength of beam strong shaft，kip − in(N − mm)
χ = Subscript related to strong axis bending
y = Subscript is related to weak axis bending
ϕ_c =Pressure strength reduction factor，$\phi_c = 0.90$
ϕ_b =Bending load strength reduction factor，$\phi_b = 0.90$
Calculating the critical setback load of a non-migrating member(P_{el})

$$P_{el} = \frac{\pi^2 EI^*}{(K_1 L)^2} \tag{3}$$

Calculating the amplification factor(β_1).

C_m is conservatively take 1.0、α = 1.0(LRFD).

$$\beta_1 = \left[\frac{C_m}{\left(1-\frac{\alpha \times P_r}{P_{el}}\right)}\right] \geq 1 \tag{4}$$

Calculating the bending member without the lateral displacement member.

$$M_u = \beta_1 \times M_{nt} \tag{5}$$

$M_{nt}$is the required flexural strength of the component when there is no lateral displacement of the component.

$$M_{ntw} = P_r \times z_D \tag{6}$$

$$M_{ntz} = P_r \times w_D \tag{7}$$

In this case, the target search function in the Excel software is used to determine the axial bending strength, and the minimum value of the value is obtained. The value is the eccentric axial pressure strength (Pr) that the section can withstand. The value obtained in the equation of the force bending moment and the integral moment of the axial force of the integral member are compared with each other to determine the section control of the angle.

3. Example of Biaxial Eccentricity Strength Calculation

The angle steel is an L8×4×7/16 unequal angle steel with a length of 10 ft. After a series of complicated calculation processes, the length (ft) of the angle steel and the obtained eccentric shaft pressure intensity value are graphically presented.

Figure 2 shows the eccentric axial pressure intensity values of the strong axial distance when the fixed distance of the weak axis is 5 (in). The values presented in the Figureure are the values of the double eccentric pressure generated by the biaxial eccentricity. The curve change is different from the common round curve, but has different curve representations. It can be seen from Figure 2 that when the weak axis distance is 5 (in), when the strong axis distance is 0 (in), the intensity curve changes to a smooth curve with a maximum value of 39.15 kips, and if the strong axis distance moves to 0.2 (in), 0.4 (in), and 0.6 (in), The intensity change curve is the same as the round curve, but the intensity value is higher according to the distance. The higher the intensity value, the greater the eccentric shaft strength value is 43.48 kips and the distance is 0.4 (in) when the strong axis distance is 0.2 (in). The maximum intensity value is 48.92 kips, and the maximum eccentric axial pressure intensity value of the strong axis distance of 0.6 (in) is 52.96 kips, which is also the maximum value of the biaxial eccentric pressure intensity value in the study.

In the study, the farther the distance is 0.6 (in), the distance will be a limit. If the distance continues to increase, the resulting eccentric axial pressure intensity will gradually become smaller, which can be 0.8 (in), 1 (in), 2 (in), the curve change is known, and the changes of the three intensity curves are different from the other curves. The three curves are different when the length of the angle is 5 (ft) and the other intensity curves in the study. When the distance is 0.8 (in), when the angle is 11 ft, the eccentric axial pressure intensity value will be higher than the 0.6 (in) eccentric axial pressure intensity value. At this time, the eccentric shaft pressure intensity value is 34.80 kips, and when the strong axis The distance is 0.8 (in), and the length of the 13 (ft) is higher than the 0.6 (in) eccentric axial pressure intensity value, and the eccentric axial pressure intensity value is 28.50 kips.

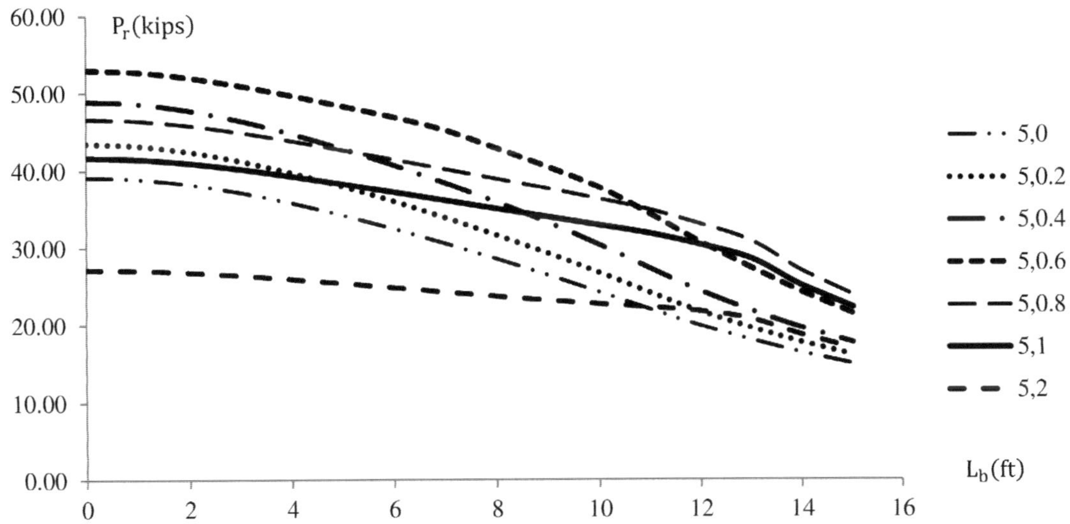

Figure 2. Biaxial eccentric pressure intensity curve

4. Conclusion

When calculating the pressure strength of the eccentric shaft, the angle steel needs to first determine which kind of limb it belongs to, considering whether it is necessary to correct the formula, to avoid the local frustration of the component, or to be destroyed when the state is not reached, when using the shaft When the force bending moment interaction equation is complicated, it is complicated and requires multiple calculations and corrections, so that the obtained value is the final accurate value. This study takes L 8×4×7/16 length 10 ft unequal angle steel as Calculation basis and analysis.

It can be seen from the calculation of the example that the calculation formula is very complicated. Therefore, when analyzing the eccentric axial pressure intensity value of the angle steel, a series of analysis and planning have been done, and the analysis steps are summarized step by step to facilitate subsequent calculation. Actions, and use the Excel target search table to analyze the calculated values.

The research results are as follows：

1. Analysis of the properties of angle steel, the various properties and parameters required for angle steel are obtained, and the calculation of the eccentric strength is started.
2. Use the calculation formula of AISC2017 specification to find the value, use Excel's target search function to bring in the formula, find the eccentric axial pressure intensity of each point, compare it with the standard value, and repeat the correction to get the exact value.
3. Use the modified formula to analyze and find the distance between the angled steel core and the strong and weak axes at each point in the study. When the value is larger, the distance is farther. After obtaining the distance parameter, the points can be calculated. The value of the eccentric shaft pressure strength.
4. When the uniaxial eccentricity, the resulting eccentric axial pressure intensity value will show a smooth curve, and the farther the distance, the lower the eccentric axial pressure strength will be; the two-axis eccentric curve will grow in the opposite direction, first by the distance 0 (in) The starting strength value is increased more, but when the distance is 0.6 (in), it is the limit, and then it will gradually decrease.
5. When the weak axis of the biaxial eccentricity is fixed at 5 (in) and the strong axis distance is 0.8 (in), the eccentric axis pressure strength value will be higher than the 0.6 (in) eccentric axis pressure strength when the angle is 11 ft. Value, the eccentric axis pressure intensity value is 34.80 kips at this time, and when the strong axis distance is 0.8 (in), the length is 13 (ft), which is higher than the eccentric axis pressure intensity value of 0.6 (in), and the eccentricity at this time The axial pressure strength value is 28.50 kips.

5. Reference

[1] American Institute of Steel Construction The AISC 2017 Steel Construction Manual, 15th Ed. and 2015 Specification Seminar(Trans Tech Publications, USA 2015)

[2] Yi Liu and Sabine Chantel 2011 Experimental Study of Steel Single Unequal-Leg Angles Under Eccentric *Compression Journal of Constructional Steel Research* Vol. 67 pp 919-928

[3] Madugula M.K.S. and Adluri S.M.R. 1994 Research and development needs in steel angle design structure. Stability Res pp. 137-148.

[4] Adluri S M R and Madugula M K S 1996 Development of Column Curve for Steel Angles Journal of Structural Engineering Vol. 122 pp 318-325

[5] Bhilawe Jyoti 2018 Experimental Study of Eccentrically Loaded Single Angle Compression Member Journal of The Institution of *Engineers (India)* Series A Vol.99 (1) pp 45-53

[6] Yuan W B , Zhan W , Xu J and Yu Nan ting 2016 Nonlinear instability of angle section beams under uniformly distributed loads *International Journal of Steel Structures* Vol.16(2), pp 309-315

[7] Hussain A, Liu Y P and Chan S L 2018 Finite Element Modeling and Design of Single Angle Member under Bi-axial Bending Structures November Vol.16 pp 373-389

[8] Kim C S, Park H G, Lee H J, Choi I R and Chung K S 2017 Eccentric Axial Load Test for High-Strength Composite Columns of Various Sectional ConFigureurations Journal of Structural Engineering Vol.143 (8) p.04017075

[9] Bradford M L X 2016 Flexural-torsional buckling of high-strength steel beams. Journal of Constructional Steel Research 124 pp 122–31

[10] Ban H S Shi Y Wang Y. 2012 Residual stress tests of high-strength steel equal angles. Journal of Structural Engineering 138(12) 1446–54

[11] Shi G, Liu Z, Ban H, Zhang Y, Shi Y and Wang Y 2012 Tests and finite element analysis on the local buckling of 420 MPa steel equal angle columns under axial compression *Steel & Composite structures* 12(1):31–51.

IOP Conf. Series: Materials Science and Engineering **938** (2020) 012007 doi:10.1088/1757-899X/938/1/012007

Effect of the Internal Pressure and External Loads on Nozzles in Cylindrical Vessel

Murat Bozkurt*, David Nash and Asraf Uzzaman

Department of Mechanical and Aerospace Engineering, University of Strathclyde, 75 Montrose Street, G1 1XJ, Glasgow, UK

murat.bozkurt@strath.ac.uk

Abstract. Understanding the potential for supporting the maximum loading conditions in the system is a key feature in the design and analysis of pressure vessel applications. This is especially important for thin-walled pressure vessels, when stresses even reaching the initial material yield point could lead to very dangerous situations. Pressure vessels may be subjected to stresses arising from a variety of loading conditions including internal pressure and multiple external loads from attached piping systems. Once the yield point has been exceeded, the structure can accommodate more loading until the plastic zone becomes excessive leading to plastic collapse. This can be challenging to establish especially when external loads act in tandem with internal pressure. Therefore, this paper develops a finite element method for the limit load analysis of a single-nozzle cylindrical pressure vessel under internal pressure and external loading in a variety of combinations. Thereafter, a parametric study is presented covering various loading conditions, both singly and in combination. Finally, a comparison is made shown the interaction effects of the effects on the limit load for changes in vessel geometry and appropriate conclusions drawn.

1. Introduction

Cylindrical structures with transverse openings are often used in pressure vessel technology. Local stresses occur especially in these intersection regions due to external loads and internal pressure. There are many theoretical solutions in the literature for the determination of local stresses which use shell theory and also finite element approaches. Xue et al [1] argue that despite the various analytical approaches and mathematical solutions to various problems in the literature, development is still needed for branch junction procedures. Xue et al. provide theoretical solutions related to external loading for two intersecting cylindrical structures. The proposed approaches are in agreement with the finite element method and WRC Bulletin 297 [2] which is a thin shell based solution, widely used in industry, but only provides shell-type stresses in the vessel and nozzle.

In pressure vessels with nozzle connection, the nozzle shell connection areas are the areas where the weakest points are due to the presence of stress concentrations arising from the intersection cut. The radius ratios d/D are important and shell theory is reliable for d/D <0.3 but has weaknesses when this rises to 1.0. It is noted that much of the literature is based on establishing the elastic stress solution with little available on limit load calculation for multiple load combinations. Sang et al. [3], attempted to obtain the results of inelastic stress analysis of a pressure vessel and nozzle connection with radius ratios of 0.526 (d/D = 0.526). These results were solved under increasing internal pressures with the help of experimental studies and nonlinear finite element methods (FEM). The analyses were supported by experimental studies to demonstrate limit loads originating from internal pressure and to compare with the results of the FEM. In addition, a burst testing was carried out to understand and validate the limit design method. In addition, the authors stated that this method does not require the elastic analysis stress categorization procedure. They also signified that the applicability of method was much simpler than the elastic design procedure.

In other studies, Mackenzie and Li [4] proposed an alternative solution to find the permissible plastic load in the pressure vessel design with the analysis approach. A three-dimensional finite element analysis was performed for cantilever beam in bending and nozzle combinations to examine the problem. Then internal pressure, torque and combined loads were applied on a cylindrical vessel with a single nozzle. TES (Twice Elastic Slope) Criterion, PW (Plastic Work) Criterion and TI (Tangent Intersection) Criterion were used to compare the proposed approach. As a result of the study, when the material shows strain hardening, higher plastic load values were obtained according to the currently available methods.

The 'elastic compensation method' used by Nadarajah et al. [5] was used for internal pressure and moment loading conditions in nozzle cylinder connections. With this method, limit and shakedown intersection diagrams can be obtained. In addition, the mentioned method was used to determine the upper and lower limit of the nozzle cylinder junction. The main objective is to investigate a highly complex vessel problem for the nozzle-cylinder intersection under the internal pressure and the nozzle subjected to torque loading. In conclusion, it is stated that the obtained results were compared with the existing methods and better values were obtained.

Furthermore, Dekkert and Stikvoort [6] conducted research on the stress intensities for the nozzle - cylinder connection. After reviewing the existing standards, some standards such as the PD5500[7], the German AD-Merkblatter [8] and the Dutch Rules for Pressure Vessel have been suggested that it is unreliable for thin-walled and relatively large nozzles. Thereupon the authors determined the purpose of the study as to provide the significant differences between the existing design codes and standards when calculating the stress intensity occurring at the nozzle-cylinder junctions. For this purpose, they compared the results using the existing design codes and finite element method. The proposed analyses have allowed for a more reliable calculation of the external loads on the nozzle.

As such, in this study, a finite element approach is used to develop a suitable model and undertake a parametric limit loads study for cylinder - cylinder intersections subject to internal pressure and combinations of external loads which will lead to an improved design approach.

2. Statement of the Problem

Establishing the maximum loading conditions to be applied in any pressure vessel is of great importance. Traditional design sets a limit on the membrane and membrane plus bending stresses but it is preferable to establish the limit load and thereafter apply a suitable factor of safety. It is desirable to establish the limit load for the system and to thereafter apply a suitable factor of safety. By means of proving the technique, simple closed form limit load calculations for a perfect cylinder are used to verify the FEM approach. In their previous work, Bozkurt et al. [9] proposed a high-fidelity model approach for the cylinder nozzle combination exposed to external loads in the elastic region. As a continuation of this study, a parametric limit load study will be introduced as described in the following steps.

3. Finite Element Model

The finite element program ANSYS [10] was used to simulate the limit load analysis for internal pressure and external loading. An accurate and reliable FEM for nozzle - cylinder interactions has been presented and the details of the FEM approach have also been described in Bozkurt et al [9]. SA-516 Grade 70 Carbon steel pressure vessel plates were selected as the shell material in the analyses to be performed. The material has a Young modulus of 200GPa and a Poisson ratio of 0.29. ASTM A266 Grade 2 Carbon Steel was selected for the nozzle material. Young modulus of the nozzle material was 190GPa and Poisson ratio was 0.29. Yield stress of the shell and nozzle were 260MPa and 290MPa respectively.

In addition, a high fidelity model approach has been introduced with a detailed parametric study on Finite Element Modelling [9]. In the validation study on the model, very close results were obtained with studies in the literature (approximately 1% difference). In addition, parametric results obtained with the same model were compared with WRC 537[11] good agreement achieved.

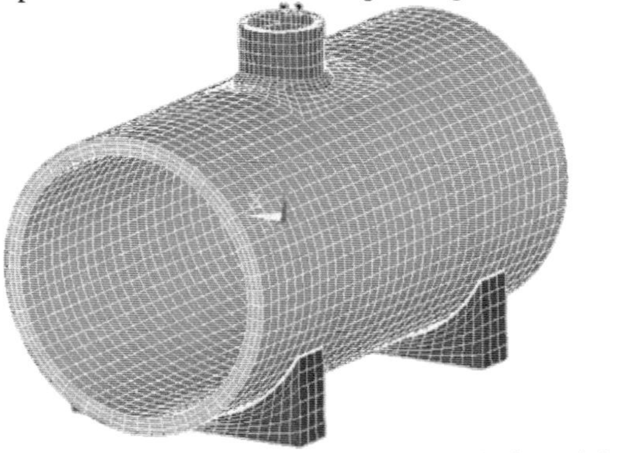

Figure 1. Perspective view of the meshed model

4. Verification of Finite Element Model

While the verification of the model was carried out, the results obtained from the finite element method as well as mathematical methods were used. These results were obtained in 3 stages. Firstly, the maximum internal pressure that the model can carry is examined. Then, the maximum tensile loading status was examined. Finally, the moment effect that will occur as a result of local loads on the nozzle is taken into consideration. The verifications are explained in detail below. For each validation study, the yield stress (σ_y) of the shell was 260MPa, the inner radius of the shell was 250 mm and shell thickness was 33.5mm. Also, the yield stress (σ_y) of the nozzle was 290MPa, the inner radius of the shell was 62.5mm and nozzle thickness was 15mm.

4.1. Verification of limit load for internal pressure and external loading applications

This verification study was carried out only for a single nozzle cylindrical vessel subjected to internal pressure. As can be seen in the Figure 1 an open end model was used and the movement was restricted to the bottom of the saddle. The friction force between the saddle and the nozzle was determined as 0.45. Taking into account the Perfect Plasticity and Tresca criteria, the force and limit required for the initial yield to be realized can be calculated by the following equation (1) and equation (2). Since the yield point of the shell is smaller than the nozzle, the calculations will be made to obtain the limit load in the shell.

$$\text{First yield:} \quad p_y = \frac{\sigma_y}{2}\left(1 - \left(\frac{r_i}{r_0}\right)^2\right). \tag{1}$$

$$\text{Limit load:} \quad p_l = \sigma_y \ln\left(\frac{r_i}{r_0}\right). \tag{2}$$

Table 1. Comparison the limit load results for internal pressure and external loading

Loading Type	Hand Calculations	FE Results	Difference
Internal Pressure ($P_{int\text{-}max}$)	32.604MPa	30.333MPa	**6.90%**
Moment Effect (Vx_{max})	6773.17kN	6640kN	**1.96%**
Tensile Loading (P_{max})	19132kN	19333kN	**1.04%**

The difference between the results obtained from the finite element model and the mathematical results that can be seen in the table is approximately 2%

5. Parametric Studies

In the parametric studies, a single-nozzle pressure vessel fitted on the saddles will be used. Nozzle and shell sizes will vary according to the parameters to be applied on the structure. The model is restricted to the movement in each direction with the boundary conditions under the saddle. Internal pressure and external loading conditions are the variables to be used in the studies. A cylindrical pressure vessel with a single nozzle connection was used for parameter studies. Maximum internal pressure or local loads will be analysed for different cases. The local loads to be applied are represented in Figure 2.

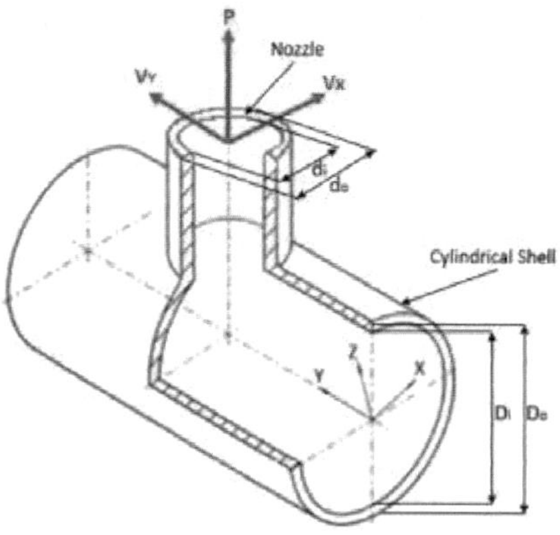

Figure 2. Load conventions for the model

5.1. Loading cases

5.1.1. 1st Case - Changing nozzle thickness. In this study, cylinder sizes for 1 nozzle cylinder combination remained constant for all analyses (D_i= 500mm, D_o = 530mm). Nozzle internal diameter (d_i) is fixed at 125mm. Nozzle thickness (t) is increased by 5 mm in each analysis. In the first case, there is no external load apply to the FEM. In the other case, the external loads affect the system from the nozzle centre. There is a 100kN loading at the same time in X, Y and Z directions. The structure is exposed to combined external loads and it is intended to be found with the limit internal pressure finite element approach. In both cases, an increase in maximum internal pressure values that can be applied in each 10 mm thickness increase was observed in a similar manner.

When the results are examined, the increase in wall thickness caused an increase in the limit pressure values in both cases. If both cases compare within themselves, the limit load is higher only in the case of 'no external load' at 5mm wall thickness, while the limit pressure is higher in the external load application after 15mm. As a result, it can be said that the effect of external loading on the limit load values is less at low thicknesses. In addition, as can be seen in the graphic on the right, the difference for both cases is approximately 4 percent for each thickness.

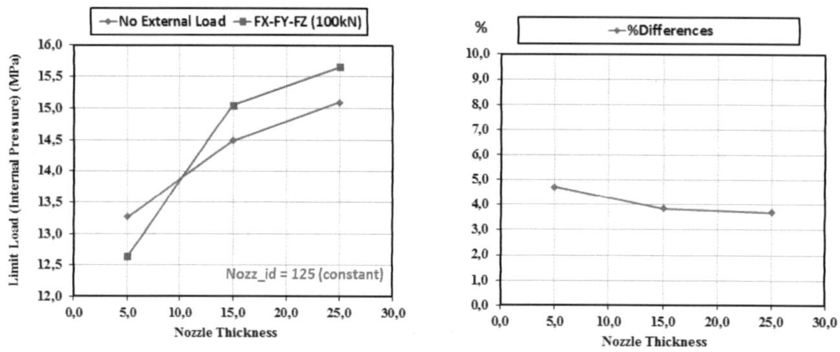

Figure 3. Comparison of the limit loads for changing nozzle thickness

5.1.2. 2nd Case - Changing shell thickness. In this study, the nozzle sizes were kept constant for the same model (d_i = 125mm, d_o = 155mm). Cylinder thickness can be seen from the Figure 4. as well as 5 mm increased in each analysis. Similar to the previous study, the internal pressure was determined by FEA as the maximum resistance of the model which was not exposed to any external load.

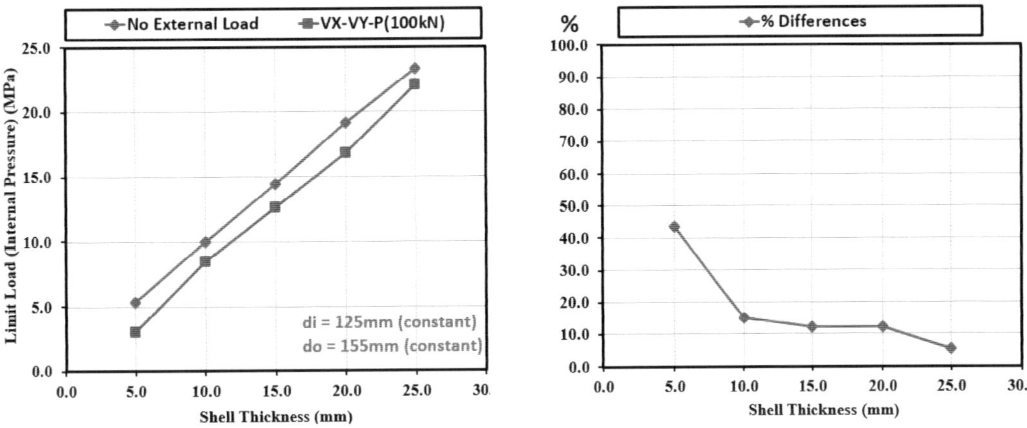

Figure 4. Comparison of the limit loads for changing shell thickness

Subsequently, 100kN loading was performed in 3 directions (Vx, VY, P) at the same time to be nozzle centred. In both cases, as the thickness of the shell increases, the strength of the structure increases and the limit increases linearly. In the case of external loading, the limit will be even less for the internal pressure with the effect of local loads. This difference is approximately 45% for 5mm wall thickness. As the nozzle thickness increases, the limit pressure difference decreases gradually. Values for each case are given in the Figure 4.

5.1.3. 3rd Case - Applying external loading for only 1 direction. For this study, internal pressure values are used as constant values. First of all, a model that was not exposed to internal pressure was used. Then, the internal pressure was increased by 5MPa in each analysis. All analyses were performed with the same dimensions. A single external load is used for each case. Limit external load values are independent of each other for X, Y and Z directions.

In pressure vessels, the internal pressure leads to circumferential and longitudinal stresses on the shell. Likewise, the radial tensile load (P) applied in the Z direction contributes to the longitudinal and circumferential stresses on the shell. Stress calculations due to radial tensile loading are given in detail in WRC537 Bulletin. When these calculations in the Bulletin are taken into consideration, as the internal pressure increases, the radial load will decrease in each step since the limit will be approached further. As can be seen from the Figure 5, the radial load decreased by approximately 30% for the 20MPa internal pressure increased.

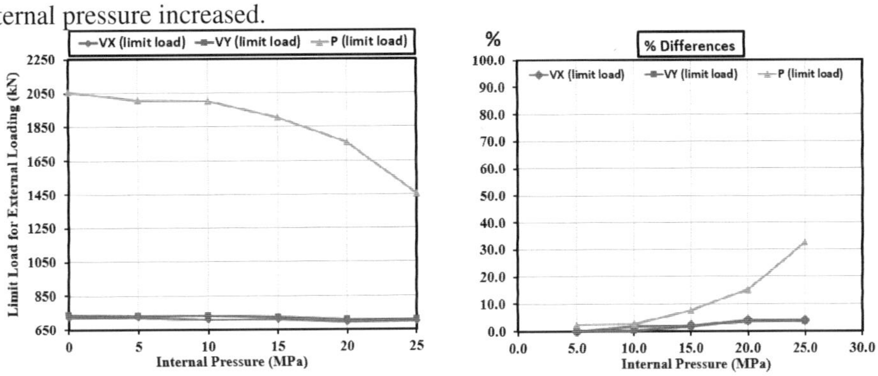

Figure 5. Comparison of limit loads for external load applied separately in 1 direction

5.1.4. 4th Case - Applying combined external loading. This study is very similar to the method described in Case 3, with the only difference is that combined loading from 3 directions are applied to the centre of the nozzle at the same time. The external load is fixed and 100kN for each direction.

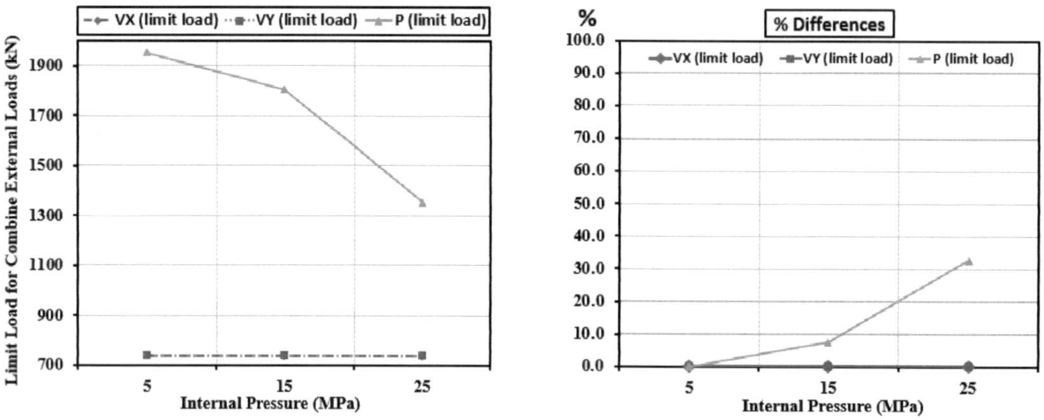

Figure 6. Comparison of limit loads for external load applied together all in 3 directions

The results of the FEA obtained for internal pressure are shown in the Figure 6. The maximum limit pressure differences for VX and VY loadings are approximately 1%, but the difference for tensile limit loads (P) is approximately 32%.

As a result, very similar results were obtained with Case 3. If the 3rd and 4th cases are compared, it can be said that a more minimal difference is observed when the longitudinal and circumferential shear forces are applied together. In other words, changing in the internal pressure are almost unaffected by the combined limit shear forces.

6. Results and Discussions

The parametric study was completed in four main stages. Firstly, the thickness of the nozzle is kept constant and the maximum applied internal pressure value is determined. In addition, the same analyses were repeated with the effect of combined external forces on the nozzle.

In the second case, with a similar approach, dimensional changes were made on the main body by keeping the nozzle dimensions constant. The thickness of the cylinder was increased by 5mm in each analysis. As can be seen in the limit pressure calculations in pressure vessels, the most important variable is the inner and outer diameter ratio. As a result of this approach, when the thickness increases, the limit load increases linearly.

In Case 3, the internal pressure values were increased by 5MPa in each analysis and limit loads were determined for external loading conditions. In the stress calculations of shear force, Bijlaard's [12] calculation approaches were shown in WRC537. According to this approach, the shear forces acting on the nozzle do not cause longitudinal and circumferential stresses in the shell. Therefore, the increase in pressure inside the vessel reduces the shear limit forces to a very small extent (approx. 3%). As a result of this analysis, it was observed that the longitudinal (V_L) and circumferential (V_C) shear forces transfer the material to the plastic region at much lower values. On the Z axis, tensile loading (P) was involved. Tensile loading directly affects the longitudinal and circumferential stresses on the shell [2]. For this reason, to transfer this material into the plastic zone, approximately 3 times more load should be applied. Therefore, tensile limit load values tend to decrease as the internal pressure increases. These results were consistent with the current approach.

Finally, in Case 4, a study similar to Case 3 was conducted. A loading was performed on the nozzle in 3 axes. For each loading state, the loading values were 100kN for 2 axes and the internal pressure value is fixed. Limit load value was tried to be found for the other axis. As a result, the biggest difference was again with tensile loading of about 32 percent. The results were obtained in a manner similar to the case 3 and are quite consistent with the existing approaches.

7. Conclusions

As a result of this study, a comprehensive parametric study was obtained for a nozzle - cylinder combination which was subjected to internal pressure and external loading conditions. By comparing the existing mathematical calculations with the finite element method, the reliability of the results has been demonstrated. In further studies, additional modifications such as reinforcement plates can be added to the existing model to enhance the results. Limit load analyses can also be performed related to cyclic loads. In the end of all these studies, it is aimed to develop a universal finite element analysis approach leading to a closed form set of equations that can resolve many cylinder-cylinder combinations.

Acknowledgements

The first author was supported by the Republic of Turkey Ministry of National Education (MoNE-1416/YLSY).

8. References

[1] M.D. Xue, D.F. Li, and K.C. Hwang, (2005). Theoretical stress analysis of intersecting cylindrical shells subjected to external loads transmitted through branch pipes. *International Journal of Solids and Structures*, 42(11-12), 3299-3319.

[2] J.L. Mershon, K. Mokhtarian, G.V. Ranjan, and E.C. Rodabaugh, "Local Stresses in Cylindrical Shells Due to External Loadings On Nozzles-Supplement to WRC Bulletin No. 107 (Revision I)" *Welding Research Council Bulletin No. 297*. (1984)

[3] Z.A.E. Sang, L.P. Xue, Y.J. Lin and G.E.O. Widera, (2002). Limit and burst pressures for a cylindrical shell intersection with intermediate diameter ratio. *International Journal of Pressure Vessels and Piping*, 79(5), 341-349.

[4] D. Mackenzie, and H. Li, (2005). A plastic load criterion for inelastic design by analysis.

[5] C. Nadarajah, D. Mackenzie, and J.T. Boyle, (1996). Limit and shakedown analysis of nozzle/cylinder intersections under internal pressure and in-plane moment loading. *International Journal of Pressure Vessels and Piping*, 68(3), 261-272.

[6] C.J. Dekker and W.J. Stikvoort, (1997). Pressure stress intensity at nozzles on cylindrical vessels: a comparison of calculation methods. *International Journal of Pressure Vessels and Piping*, 74(2), 121-128.

[7] BS 5500:1991, "G.2.2., Radial loads on cylindrical shells, G.2.3 -External moments applied to cylindrical shells." *British Standards Institution*, London, (1991)

[8] AD-Merkblatter, Taschenbuch Ausgabe 1997. Verband der Technischen Uberwachungs- Vereine e.V., Essen (Germany), 1997.

[9] M. Bozkurt, D. Nash and A. Uzzaman, "Investigation of the Stresses and Interaction Effects of Nozzle-Cylinder Intersections When Subject to Multiple External Loads" *ASME 2019 Pressure Vessels & Piping Conference. American Society of Mechanical Engineers Digital Collection.*

[10] ANSYS v19, Finite Element Code, *ANSYS Inc*, 2019

[11] K.R. Wichman, A.G. Hopper and J.L. Mershon, "Precision Equations and Enhanced Diagrams for Local Stresses in Spherical and Cylindrical Shells Due to External Loadings for Implementation of WRC Bulletin 107" *Welding Research Council Bulletin No. 537*. (2013)

[12] Bijlaard, P. P., "On the Effect of Tangential Loads on Cylindrical and Spherical Shells," Unpublished, available in the files of PVRC, *Welding Research Council*

IOP Conf. Series: Materials Science and Engineering **938** (2020) 012008 doi:10.1088/1757-899X/938/1/012008

Real-time vision-based grasping randomly placed object by low-cost robotic arm using surf algorithm

A Beyhan[1], N G Adar[1]

[1] Bursa Technical University, Mechatronics Engineering Department, Bursa, Turkey

gokhan.adar@btu.edu.tr

Abstract. Vision-Based Manipulation is popular and still have open issues in robotics. The camera is a very important part of this method to obtain desired data with image processing techniques. In this study, the Dynamic position-based look and move method was selected to control the 4 DOF robotic arm. For this method, the Kinect camera was used for image processing. Kinect is a special camera which consists of both RGB and infrared camera. SURF algorithm was selected to detect a target object from the target scene using Kinect RGB camera. 3-D target object localization was calculated using Kinect infrared camera with the point cloud. The obtained target object's location is according to the camera and transformed according to the robotic arm base. Using inverse kinematics, desired joint angles were calculated according to the object position. Therefore, the robot is provided to make the desired motion and grasping the object by using gripper. Real-time implementation of the proposed method is carried out using Matlab-Simulink.

1. Introduction

Robotics is one of the most important topics of today's engineering in Industry 4.0. Because today, robots are used in a wide area from robotic arms to humanoid robots. Various cameras are used in studies on robotic arms. These cameras act as sensors for the robotic arm. The data obtained from the cameras with image processing techniques are used in the control of the robot arm. Control of the robotic arms using the data obtained from the cameras is called Visual Servoing. It is divided into four main groups [1].

Visual servo control consists of two parts: image processing and robot control. In the image processing section, the desired data is obtained from the image space. In the control part, the robot is controlled by using this data.

There are many studies on image processing techniques. The purpose of these studies is to detect the target object in the scene. Local feature detectors are used to extract the features in the image and are divided into four categories as edge, contour, corner, and region [2].

Speeded Up Robust Features (SURF) is one of the most powerful and popular image processing algorithms to detect targets [3]. It is a robust local feature detector and an improved version of the Scale Invariant Feature Transform (SIFT) algorithm [4]. There are many studies in the literature where the SIFT algorithm is used to control robotic arms [5-11].

SURF was inspired by the SIFT algorithm. Its' principles and steps are based on the SIFT. but SURF's one of the main objectives is to make computational efficiency.

Randomly placed objects were handled using a 2D vision system used for image processing while 3D cad data is used for generating 3D coordinate. The algorithm is based on the pinhole camera model. Shaver handles were selected to recognize and was implemented using MATLAB [12].
Ali et al. used an H20 mobile robot that has dual arms where each arm consists of 6 revolute joints with 6-DOF and 2-DOF grippers for grasping and placing operation for labware. The robot has a Kinect v2 sensor for image processing. SURF algorithm was used to recognize the target object. By the way forward and inverse kinematics of arm were obtained to control robot [13].
Seib [14] et al. used GiGo service robot to grasp tasks. Gigo was built with a robotic arm and a Kinect camera to recognize the target object by using the SURF algorithm.
Suligoj et al. used a multiagent robot system which is composed of two Fanuc robots. One of these robots was equipped with a stereo camera while the other was guided for tracking, tooling, or handling operations. In the image processing stage, two methods were selected to find objects which are Hough method and SURF respectively [15].
Anh et al. aimed to grasp objects using a robotic arm with an eye in hand camera. For this purpose, Image-Based Visual Servoing was used to control the robotic arm. To identify the target object SURF algorithm was selected [16].
Hofer et al. [17] proposed a Visual Servoing method to grasp objects. For this propose they used 7-DOF iARM robotic arm and RGB-D camera with a pan-tilt mechanism. In this set up to identify the object in the scenario SURF algorithm was used. Shalfan et al. [18] studied Image-Based Visual Servoing for 7-DOF robotic arm mounted on a wheelchair. SURF algorithm was selected for Visual Servoing for both eye-in-hand and eye-to-hand camera configuration. Many studies have been made about 7-DOF robotic arm with a wheelchair for disabled people [19-21].
Khusheef et al. [22] set up experiments including a 6-DOF robotic arm, low-cost camera, and conveyor belt system. An object placed on to the conveyor belt and detected using the SURF algorithm. After detecting the object, the robotic arm moved to grasp and relocate the object.
In this study, the Dynamic position-based look and move technique was selected to control the 4-DOF robotic arm with a Kinect camera.

2. Vision-based manipulation

Visual servo control is a control method that the camera uses to make the robot perform the desired motion. In this method, the camera is used as a sensor to obtain useful information for robot motion. In the literature, there are different visual servo structures. This structure consists of two main parts: Computer vision and robotic manipulation. In the computer vision part, the target object is detected in the scene using the SURF algorithm.
In the Computer vision part, the Kinect camera is used to determine its localization (x-y-z coordinate) using point cloud. In the robotic manipulation part, kinematic analysis (Forward-Inverse kinematics) is used to control and move the robot for the target object the general block diagram is given in figure 1.

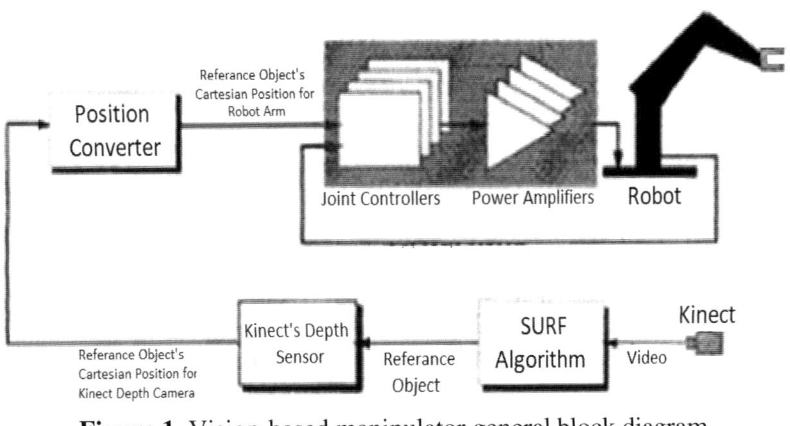

Figure 1. Vision-based manipulator general block diagram.

IOP Conf. Series: Materials Science and Engineering **938** (2020) 012008 doi:10.1088/1757-899X/938/1/012008

2.1. Vision-based object detection

2.1.1. Object detection. In the literature, different image processing algorithms for detecting the object in the image space are available. In this study, SURF which is one of the most powerful and popular image processing algorithms to detect targets algorithm was selected. It is a robust local feature detector and an improved version of the SIFT algorithm. SURF was inspired by the SIFT algorithm. Its' principles and steps are based on the SIFT but SURF's one of the main objectives is to make computational efficiency [14]. It is a scale-invariant feature detector which means that it is not affected by scaling and rotation (scale and rotation invariant detector).
To detect Interest point, SURF is used on the Hessian matrix which is as follow:

$$H = \begin{bmatrix} I_{xx}(x,\sigma_D) & I_{xy}(x,\sigma_D) \\ I_{xy}(x,\sigma_D) & I_{yy}(x,\sigma_D) \end{bmatrix} \quad (1)$$

where I_{xx}, I_{xy}, and I_{yy} are second-order Gaussian smoothed image derivatives.
In equation 1, there are second-order derivatives. These derivatives make the computation complex and take a long time. To eliminate this complexity, SURF uses of integral images method to efficiently compute the Hessian matrix.
Integral images provide a quick calculation of a given rectangular area calculating the sum of pixel values as follow:

$$I_\Sigma(x) = \sum_{i=0}^{i\leq x} \sum_{j=0}^{j\leq y} I(i,j) \quad (2)$$

Using the integral image method, The Hessian matrix is roughly approximated.
SURF uses the Fast-Hessian detector based on the determinant of the Hessian matrix. This determinant is used to find the maximum and minimum points in the images taking the second-order derivative. The maximum determinant of this matrix represents the image regions (blob). Besides, scale and position are also found using the determinant of the Hessian matrix [7]. The determinants of the Hessian matrix are described below.

$$\left|H(I(x,y))\right| = \frac{\partial^2 I}{\partial x^2}\frac{\partial^2 I}{\partial y^2} - \left(\frac{\partial^2 I}{\partial x \partial y}\right)^2 \quad (3)$$

The determinant of Hessian can be calculated using the estimated Gaussian as follows:

$$\left|H_{approx}\right| = D_{xx}D_{yy} - 0.9D_{xy}^2 \quad (4)$$

where D_{xx}, D_{yy}, D_{xy} are the second-order Gaussian derivate which is approximated based on box filters.

2.1.2. 3-D object localization. Kinect is a special camera system developed by Microsoft (Figure 2).

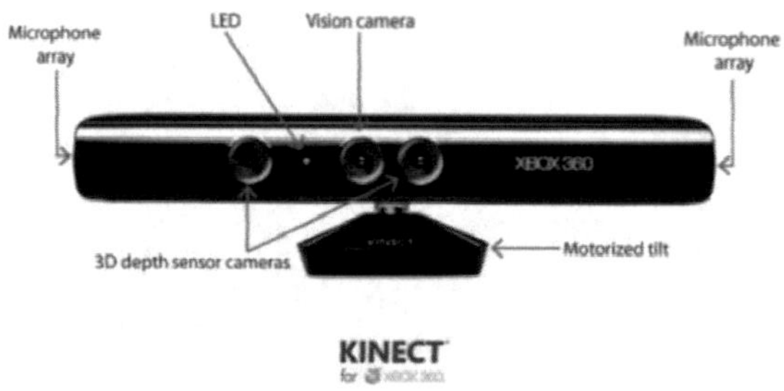

Figure 2. Kinect v1 Camera.
It consists of an infrared laser emitter, an infrared camera, and an RGB camera. The RGB camera is a classic camera that is based on three basic colors (red, green, and blue). The depth detection system (3D depth sensor camera) is formed by an IR camera and infrared projector. This structure was developed by PrimeSense. The IR laser emits structured infrared lights with a certain frequency. These IR lights hit the objects and create an observed pattern onto the scene. IR camera capture this observed pattern.

Therefore, a depth map containing the distance of an object in the scene to the camera can be extracted with these two components.

2.2. Forward and inverse kinematics of robotic arm

The details of the forward and inverse kinematics of a 4-DOF robotic arm given below have been presented in [23]. Brief information was given here.

The robotic arm is given in the Cartesian coordinate system in figure 3.

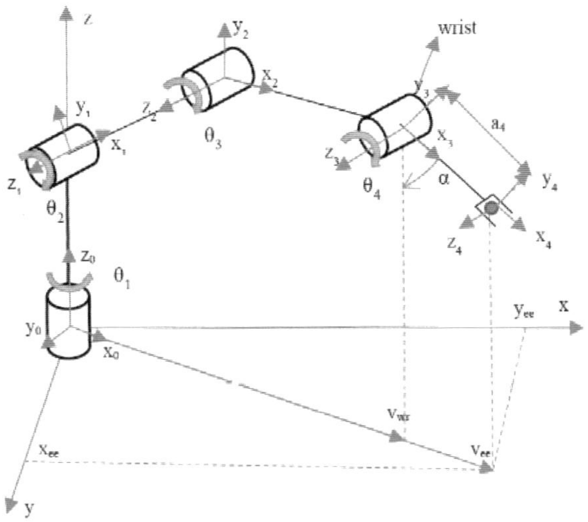

Figure 3. The coordinate frame to each link.

In figure 3 α is the wrist-roll angle, x_{ee} is x coordinate of the end effector according to the base, y_{ee} is y coordinate of the end effector according to the base, z_{ee} is z coordinate of the end effector according to the base, v_{wr} is the vector of the wrist, v_{ee} is the vector of the end effector, θ_1, θ_2, θ_3 and θ_4 are the joint angles. Forward kinematics equations are obtained by the Denavit-Hartenberg (D-H) method. Table 1 is derived according to D-H parameters is given below.

Table 1. D-H Table of Robotic Arm.

i	a_i	α_i	d_i	θ_i
1	0	90°	45 mm	θ_1
2	150 mm	0	0	θ_2
3	150 mm	0	0	θ_3
4	90 mm	0	0	θ_4

The transformation matrix from base to end-effector (T_4^0) can be obtained as a product of matrices for every joint given in equations 5:

$$T_4^0 = T_1^0.T_2^1.T_3^2.T_4^3 = \begin{bmatrix} n_x & o_x & a_x & p_x \\ n_y & o_y & a_y & p_y \\ n_z & o_z & a_z & p_z \\ 0 & 0 & 0 & 1 \end{bmatrix} \tag{5}$$

where P (P_x, P_y, P_z) is the position vector of the end-effector and n, o and a are orthogonal unit vectors that define the orientation of the end-effector frame.

In equation 6-8, the Position coordinates of end-effector are given below:

$$P_x = cos(\theta_1)\big(a_2 cos(\theta_2) + a_3 cos(\theta_2 + \theta_3) + a_4 cos(\theta_2 + \theta_3 + \theta_4)\big) \tag{6}$$

IOP Conf. Series: Materials Science and Engineering **938** (2020) 012008 doi:10.1088/1757-899X/938/1/012008

$$P_y = sin(\theta_1)\big(a_2 cos(\theta_2) + a_3 cos(\theta_2 + \theta_3) + a_4 cos(\theta_2 + \theta_3 + \theta_4)\big) \quad (7)$$

$$P_z = d_1 + a_2 sin(\theta_2) + a_3 sin(\theta_2 + \theta_3) + a_4 sin(\theta_2 + \theta_3 + \theta_4) \quad (8)$$

In this study, the geometric approach is used to obtain the inverse kinematic equations of the robot. Using this method θ_1, θ_2, θ_3, and θ_4 were calculated as follow:

$$\theta_1 = atan2(x_{ee}, y_{ee}) \quad (9)$$

$$cos\,(\theta_3) = \frac{x_{wr}^2 + y_{wr}^2 - d^2 + (z_{wr} - d_1)^2 - a_2^2 - a_3^2}{2a_2 a_3} = A \quad (10)$$

$$sin\,\theta_3 = \sqrt{1 - A^2} \quad (11)$$

$$\theta_3 = atan2(A, \pm\sqrt{1 - A^2}) \quad (12)$$

$$\theta_2 = atan2(x_{wr}^2 + y_{wr}^2, d_1) - atan2(a_2 + a_3\,cos\,(\theta_3), a_3\,sin\,(\theta_3)) \quad (13)$$

$$\theta_4 = \frac{\pi}{2} - \alpha + \theta_1 - \theta_2 \quad (14)$$

3. Experiment and results

To implement the Dynamic position-based look and move algorithm, the experimental setup was obtained. The experimental setup consists of a robotic arm, Kinect camera, and computer (Figure 4). Matlab-Simulink software is used to develop the code.

Robotic Arm has 5 servo motor (Dynamixel AX-12A motor). Four of them are used for joint motion while the other servo motor is used for gripper (Figure 5).

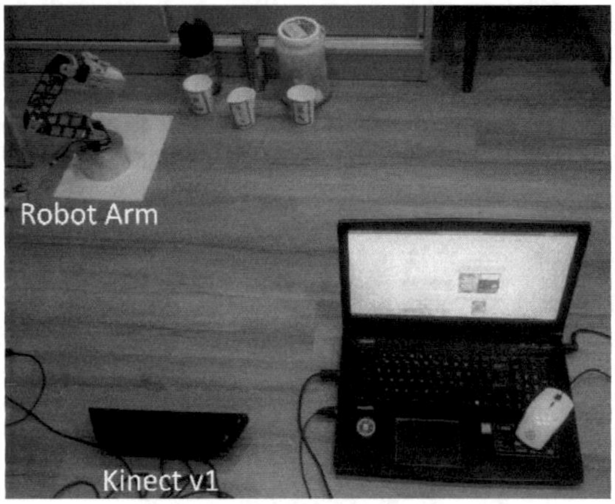

Figure 4. Experimental setup

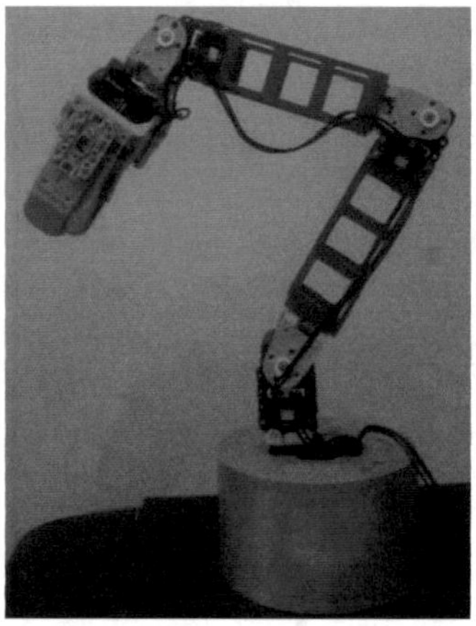

Figure 5. 4 DOF Robotic arm

The main propose of this study is that detect the object which was selected before. To detect the object, SURF which was an image processing algorithm was used. To control the robot, the kinematics of the robot was used.

The steps of the algorithm are

1- Select and upload the object.
2- Run the SURF to extract the feature points
3- Find the strongest feature points
4- Run the Kinect camera to obtain the image scene
5- Match the feature points from the target object and target image scene
6- Find the target object
7- Calculate the center point of the target object
8- Obtain the point cloud using the Kinect 3D depth camera
9- Use the target object's center coordinates to calculate the position coordinates of the target point.
10- Transfer these coordinates to the robot base coordinates

11- Calculate joint angle using Inverse Kinematics according to these coordinates

12- Stop the algorithm until the joint angle errors below 0.01°.

For experiments, a patterned cup was selected for the target object. This object was photographed and uploaded to the computer database (Figure 6).

The first step of the SURF algorithm is to obtain feature points on the target object image in the database. The feature points obtained from the target object are given in figure 7.

Figure 6. Target object

Figure 7. The strongest feature of the target object

Different objects were placed in the target image scene together with the target object. The target image scene was obtained using the Kinect RGB camera.

The strongest feature points were found in the target image scene by SURF (Figure 8).

Feature points obtained from the target object and target image scene were compared and similar ones have been matched (Figure 9).

Using the M-estimator Sample Consensus (MSAC) algorithm, it eliminates point which does not belong to the target object (Figure 10).

Figure 8. The strongest feature of the target image

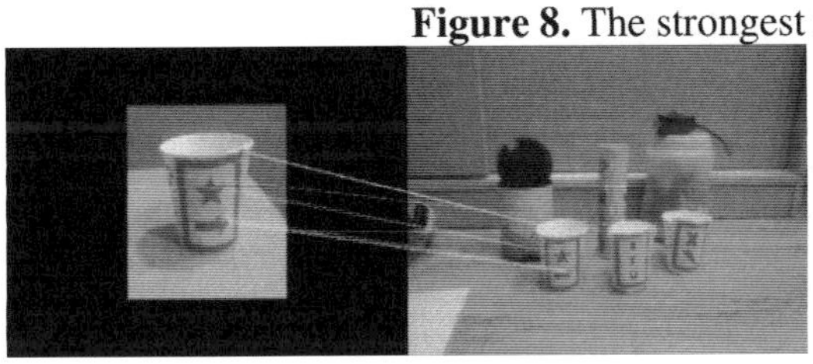

Figure 9. Matching similar points

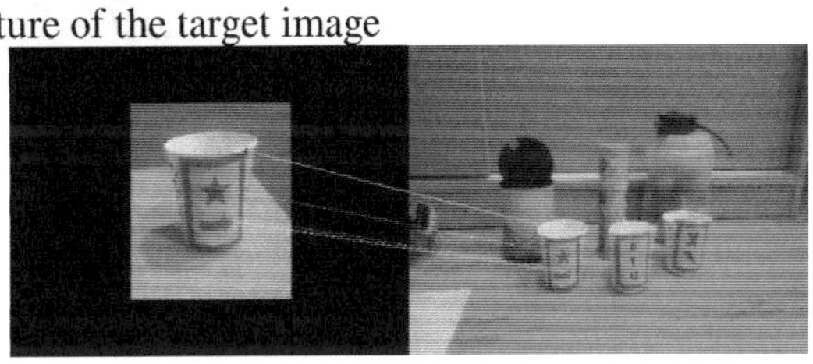

Figure 10. Elimination of unrelated point

After eliminating the unrelated point, the remaining points belong to the target object which was taken into the frame (Figure 11).

Figure 11. Assigning frame of the target object

Using this frame's corner points, the center point of the target object is calculated as follows and the center and points are showed in Figure 12:

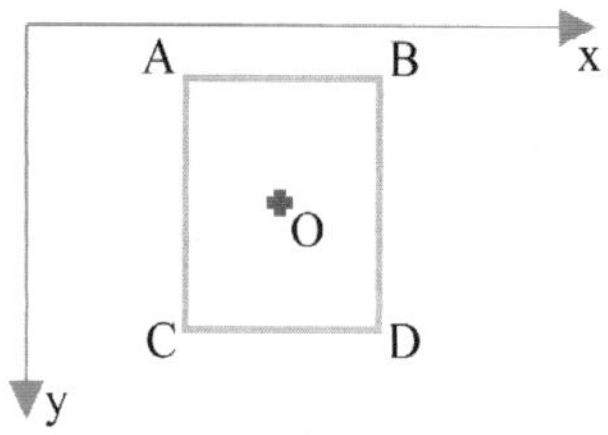

Figure 12. The center and corner point of the frame

$$O_x = A_x + \frac{D_x - A_x}{2} \tag{15}$$

$$O_y = A_y + \frac{D_y - A_y}{2} \tag{16}$$

where O_x and O_y are the center point of objects and A, B, C and D are frame's corner points.

After finding the center, the point cloud is obtained by using the Kinect 3D depth camera (Figure 13). A point cloud is a set of data points which are the x-y-z distance of the object in space. The center of the target is addressed the x-y coordinates. After having the 3D point cloud, using these addresses, the position coordinates of the object on the target scene were obtained.

The position coordinates of the target point were calculated using point cloud data with the target object's center coordinates. These target object's coordinates belong to the center of the camera. However, the position of the target object relative to the robotic arm is required. The relation between the camera center and the robotic arm's base was given in figure 14.

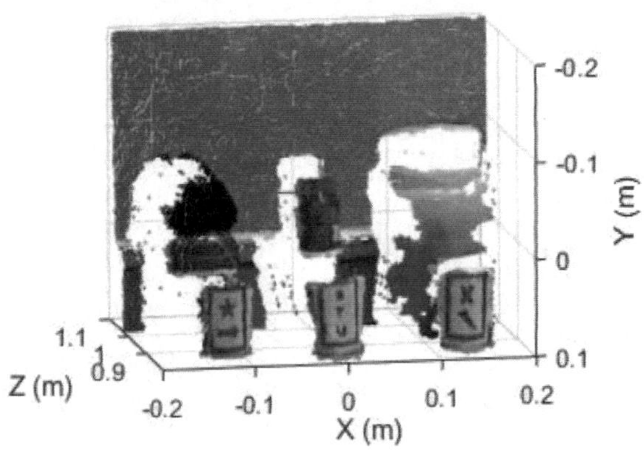

Figure 13. Point clouds of objects

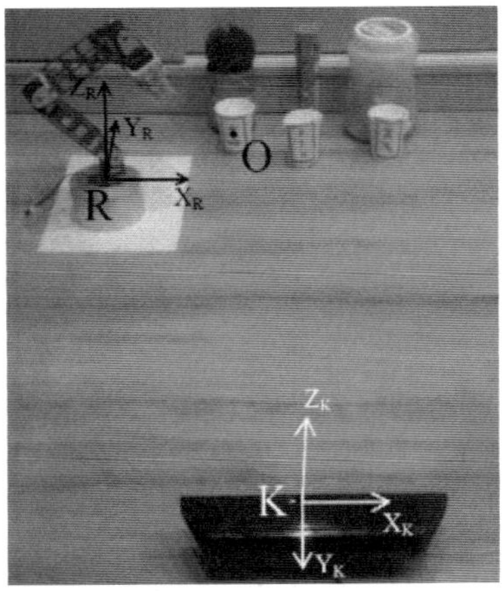

Figure 14. Image of relation Cartesian position of between Kinect and robotic arm

In figure 14, R is coordinate of the robot arm's base, K is the center coordinate of Kinect, O is the center of the object. and this position was calculated as follow:

$$ {}_{R}^{O}P_x = {}_{K}^{O}P_x + 25 \tag{17} $$

$$ {}_{R}^{O}P_y = {}_{K}^{O}P_z - 75 \tag{18} $$

$$ {}_{R}^{O}P_z = 4 - {}_{K}^{O}P_y \tag{19} $$

where ${}_{R}^{O}P$ is the Cartesian position of an object according to robot arm's base and ${}_{K}^{O}P$ is the Cartesian position of an object according to Kinect camera.

Using equation 17-19, the position of the target object relative to the robot's base has been determined. With these positions, the angle values required for the motion of the robot were calculated using Inverse Kinematics. Thus target joint angles were calculated. The robotic arm has servo motors in each joint. These servo motors have their joint angle controller. Using these angle values, the robot was moved. The robotic arm motion lasted until the angles errors were below 0.01°. After reaching the target error, the robotic arm stops to move and captured the object with its gripper (Figure 15).

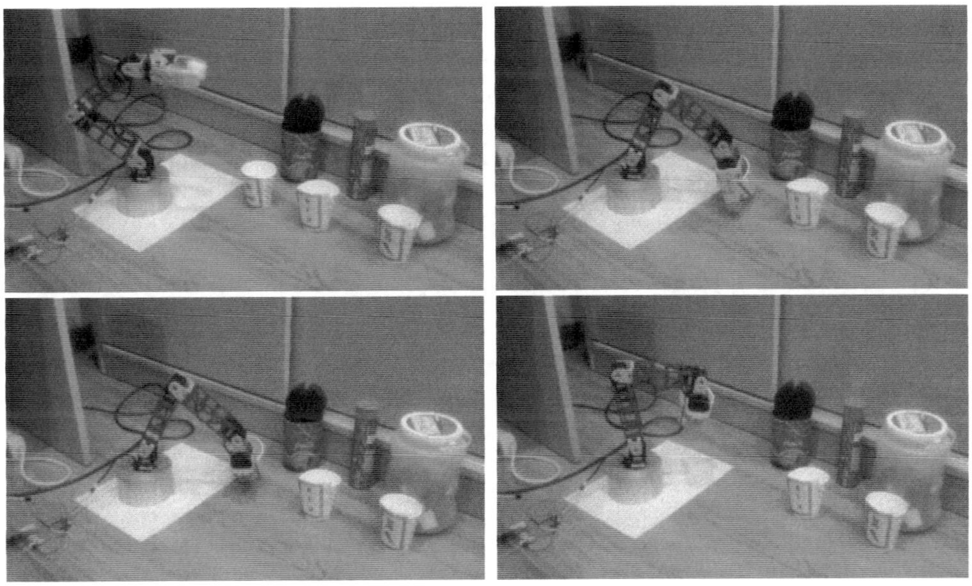

Figure 15. Move and grasp the target object

4. Conclusion

In this paper, Vison Based Robot Manipulation was implemented. 4 DOF low-cost robotic arm and Kinect camera were used for experiments. To detect the target object in the scene, the SURF algorithm was practiced with a Kinect RGB camera. the localization of the target object was found using Kinect infrared camera with the point cloud. Each joint angle was calculated using this position with Inverse kinematics. After finding the joint angles, the robotic arm was moved, and the object was grasped by the robot's gripper. The first problem of this study is the intensity of light. SURF algorithm is an image processing technique that was dependent on the intensity of light. The light intensity needs to be adjusted well to detect the target object. The second problem was the selection of the target object. The object should contain a pattern. Photographing the target object with a Kinect camera gives better results in detecting the object. The third motor of the Robotic Arm has a narrow workspace because of the mechanical limitation. Therefore, the workspace of the robotic arm has also been significantly reduced. In the future research ideas as below:

- Detecting moving objects and grasping by the gripper.
- Providing a wider workspace and designing trajectory planning for the robot.
- Assigning position of gripper according to the target object using 6 DOF robot.

Acknowledgments

This work was supported by the Scientific Research Projects Coordination Unit of Bursa Technical University. Project number: 191N008.

5. References

[1] Hutchinson S, Hager G D and Corke P I 1996 A tutorial on visual servo control *IEEE transactions on robotics and automation* **12** 651-670

[2] Salahat E and Qasaimeh M 2008 Recent advances in features extraction and description algorithms: A comprehensive survey *Proc. IEEE Int. Conf. Ind. Technol.* 1059–63

[3] Bay H, Ess A, Tuytelaars T, and Van Gool L 2008 Speeded-up robust features (SURF) *Computer vision and image understanding* **110** 346-359.

[4] Lowe D G 2004 Distinctive image features from *Int. J. Comput. Vis.* **60** 91–110

[5] Yang Y and Cao Q 2013 A fast feature points-based object tracking method for robot grasp *International Journal of Advanced Robotic Systems* **10** 170

[6] Sushkov O O and Sammut C 2011 Feature segmentation for object recognition using robot manipulation *Proceedings of Australasian Conference on Robotics and Automation*

[7] Collet A, Berenson D, Srinivasa S S and Ferguson D 2009 Object recognition and full pose registration from a single image for robotic manipulation *IEEE Int. Conf. Robot. Autom.* 48–55

[8] Chen C H and Huang H P 2013 Pose estimation for autonomous grasping with a robotic arm system *J. Chinese Inst. Eng. Trans. Chinese Inst. Eng. A/Chung-kuo K. Ch'eng Hsuch K'an*, vol. **36** 638–646

[9] Piccinini P, Prati A and Cucchiara R 2012 Real-time object detection and localization with SIFT-based clustering *Image Vis. Comput.* **30** 573–587

[10] Liefhebber F and Sijs J 2007 Vision-based control of the Manus using SIFT *IEEE 10th Int. Conf. Rehabil. Robot. ICORR'07* 854–861

[11] Patil G G 2013 Vision guided pick and place robotic arm system based on SIFT *International J. Sci. Eng. Res.* **4** 242–248

[12] Tsarouchi P, Matthaiakis S A, Michalos G, Makris S and Chryssolouris G 2016 A method for detection of randomly placed objects for robotic handling *CIRP J. Manuf. Sci. Technol.* **14** 20–27

[13] Ali M, Liu H, Stoll N and Thurow K 2017 Grasping and placing operation for labware transportation in life science laboratories using mobile robots *Adv. Sci. Technol. Eng. Syst.* **2** 1227–37

[14] Seib V, Thierfelder S, Paulus D. 2011 Object recognition tasks for service robots *8th Open German-Russian Workshop: Pattern Recognition and Image Understanding* 258-261

[15] Šuligoj F, Šekoranja B, Švaco M and Jerbic B 2014 Object tracking with a multiagent robot system and a stereo vision camera *Procedia Eng.* **69** 968–973

[16] Anh L T and Song J B 2010 Object tracking and visual servoing using features computed from local feature descriptor *ICCAS 2010 - Int. Conf. Control. Autom. Syst.* 1044–48

[17] Hofer L, Tanaka M, Tamukoh H, Nassiraei A A and Morie T 2016 Depth-based visual servoing using low-accurate arm *2016 Joint 8th International Conference on Soft Computing and Intelligent Systems (SCIS) and 17th International Symposium on Advanced Intelligent Systems (ISIS)* 524-531

[18] Elarbi-Boudihir M and Al-Shalfan K A 2013 Eye-in-hand/eye-to-hand configuration for a WMRA control based on visual servoing *2013 IEEE 11th International Workshop of Electronics, Control, Measurement, Signals and their application to Mechatronics* 1-6

[19] A. Palla, A. Frigerio, G. Meoni, and L. Fanucci, 2017 Position Based Visual Servoing control of a Wheelchair Mounter Robotic Arm using Parallel Tracking and Mapping of task objects *ICST Transactions on Ambient Systems* **4** 1–7

[20] Jiang H, Wachs J P and Duerstock B S 2013 Integrated vision-based robotic arm interface for operators with upper limb mobility impairments *IEEE Int. Conf. Rehabil. Robot.* 1–6

[21] Jiang H, Wachs J P and Duerstock B S 2014 Integrated vision-based system for efficient, semi-automated control of a robotic manipulator *Int. J. Intell. Comput. Cybern.* **7** 253–266

[22] Khusheef A S, Kothapalli G and Tolouei-rad M 2011 An approach for integration of industrial robot with vision system and simulation software **5** 18–23

[23] Beyhan A and Adar N G 2019 Detailed kinematic analysis and real time application of the 4-dof low cost robotic arm *11th International Conference on Electrical and Electronics Engineering (ELECO)* 1037-41

Practical Reduction of Harmonics for Induction Generator with Simulator Mini Micro Hydro for Educational Purposes

Syaiful Bakhri, Seflahir Dinata, M Toriqul Amien, Yanwar Saputra

Electrical engineering, Pamulang University,
Jl. Raya Puspiptek, Pamulang, Kota Tangerang Selatan, Banten, Indonesia, 15310

syaiful20201@gmail.com; dosen01138@unpam.ac.id; dosen01794@unpam.ac.id; yanwarsaputra15@yahoo.com

Abstract. Micro Hydro Power Plant as one of the power generation that relies on waterfalls or river flows as the driving force. The generator produces harmonic voltage and current waves where these waves are a distorted wave disturbance causing multiplication in the fundamental frequency. This study aims to demonstrate practical and straightforward approaches to implement the appropriate filters in reducing harmonics in an induction generator for the micro hydropower. This approach could be useful for standalone rural/remote applications with a minimum competency of an electrical engineer to arranged the hydropower as well as for educational purposes at universities. Mini custom made micro-hydro simulator is employed. Two selected filters were chosen, i.e., Band Pass Filters and High Order Passive Filters as the harmonic dampers. The Bandpass filter able to reduce voltage harmonics without load, which is THDv 0.33% to 0.26% even though it is still far in the IEEE standard, i.e. 5%, and the bandpass filter does not affect while the system connected with the load. The high order passive filter can reduce the harmonic wave with THD 67 % becomes THDi 0.003% and THDv 0.40% to 0.17%. It was demonstrated that the applicability of this filter at the simulator could be employed to adjust the harmonics for the generated output of micro-hydro.

1. Introduction

Pico and Micro Hydro Power Plant (P/MHPP) also known as Pembangkit Listrik Tenaga Micro/Pico Hydro (PLTM/PH) in Indonesia is a power generation that relies on water as the driving force for waterfalls or river flows with the capacity less 500 kW and 10 kW respectively [1]. PLTMH is also a small-scale hydropower plant, mainly because of the construction of the building and the produced power, which is not as big as a hydropower and mini-hydropower plant. This source of electricity generation in Indonesia is also promising in the future. The potency was around 143,845.30 kW, which is only 1.8% utilization[1]. In addition, considering the location of Pico and Micro-Hydro Power Plant both available in small and remote rivers, it can be very beneficial for standalone power sources[2]. These opportunities mean that a better understanding and acceptance of the public to implement this simple technology is needed. However, it is hardly found a simulator or trainer that demonstrates this technology even on a laboratory scale in Indonesia. This research uses this mini micro-hydro previously developed[3].

One of the essential components of PLTM/PH is the generator that converts mechanical energy into electrical energy. In the case of the type of generator, the induction generator draws more attention than the synchronous one, due to its lower price, better reliability, lesser maintenance, a wider range of

power selection and more straightforward starting and control [4-6]. This generator produces harmonics that are in the form of distorted disturbance waves causing multiplication for the fundamental frequency [7, 8]. As a result of these harmonics, they affect the loads, in the form of increasing the heat, which in turn causing damages. One of the simple solutions for this problem that might be applied easier in rural areas is an addition of a harmonic filter to reduce the distorted wave [7]. According to the Institute of Electrical and Electronics Engineers' IEEE, the allowable standard harmonics is around 5%. Therefore, the subject of discussion is which type of filter is the most appropriate in reducing harmonics.
Considering these realities, this study aims to demonstrate the filter approaches in reducing harmonics for an induction generator, especially for the Pico or Micro Hydro Power Plant. This practical and straightforward approach is expected to give a better understanding of the public with no particular expertise in electrical engineering to reduce the harmonics of the induction generator for their pico or micro hydropower plant arrangements. It should be noted here that a Simulator of Micro Hydro Power Plant is introduced and developed. The simulator was used both for scientific and educational purposes. Two selected filters are chosen, i.e., Band Pass Filters and High Order Passive Filters as the harmonic damper. This research analyzes the signal amplitude and the behavior of harmonics before and after filtering.

2. Theory

Induction generators are employed to convert rotary energy into electrical energy. In a self-powered induction generator, the excitation is obtained from a capacitor installed parallel to the generator output terminal [4, 5]. This type of induction generator works similarly like an induction machine in its saturation area except that there is a capacitor bank installed at the stator terminal. Because the source of the excitation of this generator comes from the capacitors mounted on the terminal, the induction machine with wound rotors or squirrel cages can be used as self-powered induction generators.
The induction generators or on all rotating machines usually produce harmonics. Harmonics are the multiplication of frequency from the fundamental frequency [4, 9]. The harmonics are usually in an odd order, causing a distorted sine wave. The effect of this distorted wave causes overheats of electrical equipment, which later may damage if it was left unmonitored.
Harmonic filters can be applied To reduce the harmonics that occur in the system. The harmonic filter functions to drain, hold, eliminate high frequencies in the system. In this study, two types of harmonic filters were used, namely. a. Band Pass Filter: a filter that passes frequencies with certain bands. b. High Order Passive Filter: a filter that absorbs frequencies at high order. The following equation is employed To determine the value of the components [10, 11]. The capacitor values are determined using Eq. 1, while the inductor values are from Eq. 2 and 3.

$$C = 1 / (2.\pi.f.X_C) \quad (1)$$

$$X_L = X_C / n^2 \quad (2)$$

$$L = X_L / (2.\pi.f) \quad (3)$$

Where C is the capacitance in Farads, L Inductance in Henry, f frequency in Hz, XC capacitive reactance, XL inductive reactance, and the n is the ratio of capacitive and inductive reactance. The amount of Total Harmonic Distortion (THD) is given in Eq. 5.

$$\text{Distortion (total dB)} = 10\log(10^{dB1/10}+10^{dB2/10}+10^{dB3/10}+...+10^{dBn/10}) \quad (4)$$

$$\text{THD (\%)} = 100 \times 10^{\text{total dB}/10} \quad (5)$$

3. Methodology

This research developed a mini simulator for Micro/Pico Hydro Power Plant. The entire block diagram of the custom made simulator can be seen in Fig. 1[3].

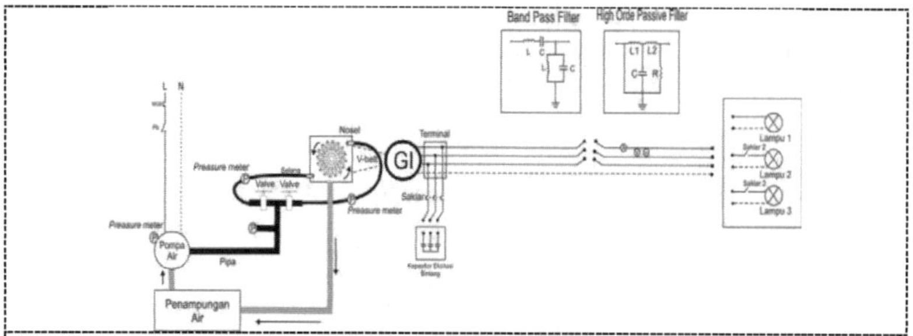

Figure 1. The custom made simulator of micro-hydro generation

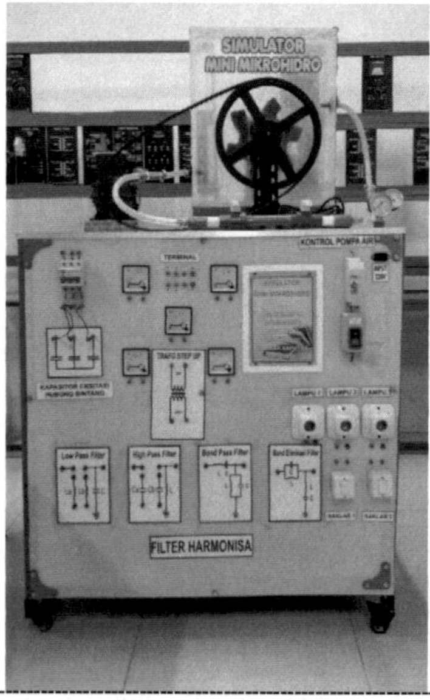

Figure 2. The final product of micro-hydro generation simulator

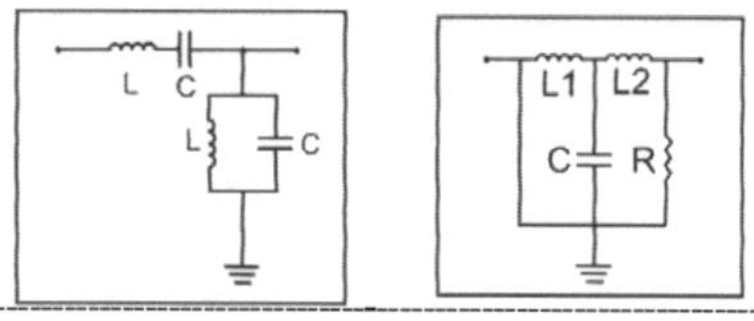

Figure 3. The applied filter, bandpass filter (left) and high order passive filter (right)

Fig. 1 and Fig. 2 show an induction generator coupled with a Pelton Turbine [12] as the driver, using the v-belt. The excitation of the induction generator circuit is arranged with a star connection system. After the excitation system is installed, the micro-hydro system is coupled with various filters, as Given in Fig.3. Testing is done to determine the characteristics of the parameters of the main simulator before and after the filter was applied.

IOP Conf. Series: Materials Science and Engineering **938** (2020) 012009 doi:10.1088/1757-899X/938/1/012009

Table 1. The developed turbine parameters

Parameter	Value
Type of Turbine	Pelton
Blade Width	5.5 cm
Turbine Diameter	25 cm
Number of Blades	6 blade
Rotation Speed	646 rpm

For the measurement, NI-USB 6009, was employed. This data acquisition measures voltage, flow, and current. The harmonic detection and calculation are done on the computer using LabVIEW National Instrument.

4. Result and discussion

Following the testing parameters that have been previously determined, in this section, the results of the test are discussed. Tests have been carried out on Harmonics Filters with an LED load lamp of 220V 3 Watt using a bandpass filter. Here is the example of micro-hydro generation implementation, including the various filter applications. One of the examples was obtained at 43 volts, and 1150 RPM generator as the waveform is given in Fig. 4.

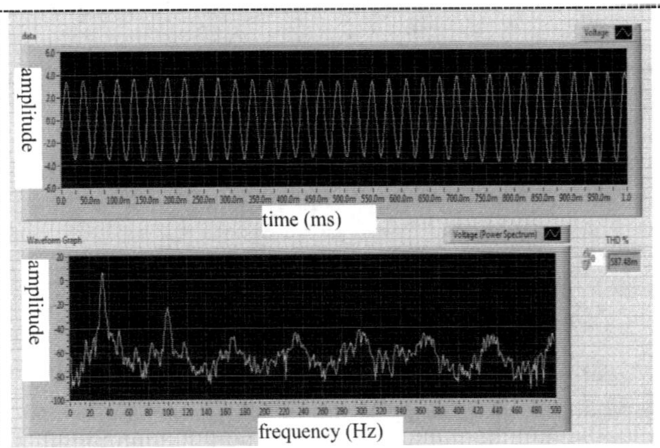

Figure 4. The voltage and frequency waveforms with a 3 Watt 220 V load before filtering

Fig. 4 shows that the sine wave voltage has a period of 0.03 seconds with a fundamental frequency at around 33 Hz with an amplitude of 8 dB. In this figure, the 3rd (100 Hz), 5th (170 Hz), 7th (233 Hz), 9th (298 Hz), 11th (366 Hz), and 13th (435 Hz) order of harmonic have amplitudes at around -24 dB, -45 dB, -45 dB, -42 dB, -45 dB and -46 dB respectively. By using Eq. 4 and 5, it was found that the THD is around 0.40 %. It clearly shows that various harmonics might produce by the induction generator and LED lights.

IOP Conf. Series: Materials Science and Engineering **938** (2020) 012009 doi:10.1088/1757-899X/938/1/012009

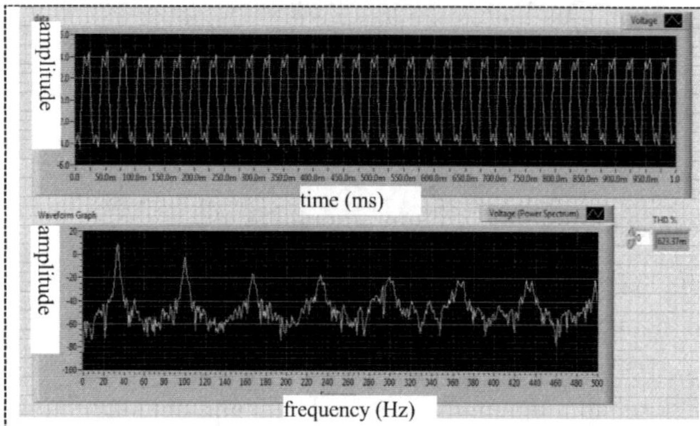

Figure 5. The current and frequency waveforms with a 3 Watt 220 V load before filtering

It can be seen in Fig. 5 the sine waveform voltage has a period of 0.03 seconds with a fundamental frequency of around 33 Hz at an amplitude of 10 dB. In Fig. 5, it can be seen that the harmonics on the 3rd (100 Hz), 5th (165 Hz), 7th (235 Hz), 9th (298 Hz), 11th (366 Hz), and 13th (435 Hz) order have amplitudes of around -2 dB, -15 dB, -17 dB, -20 dB, -21 dB, and -21 dB respectively. It can be seen that the amplitude of the current is relatively easier to be identified than the voltage.

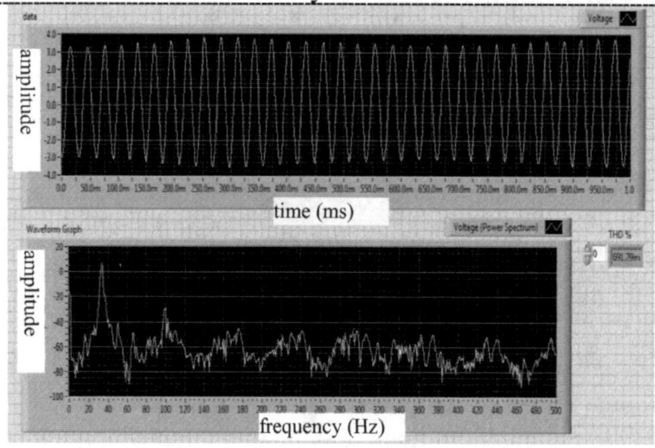

Figure 6. The voltage and frequency waveforms with a 3 Watt 220 V load after filtering

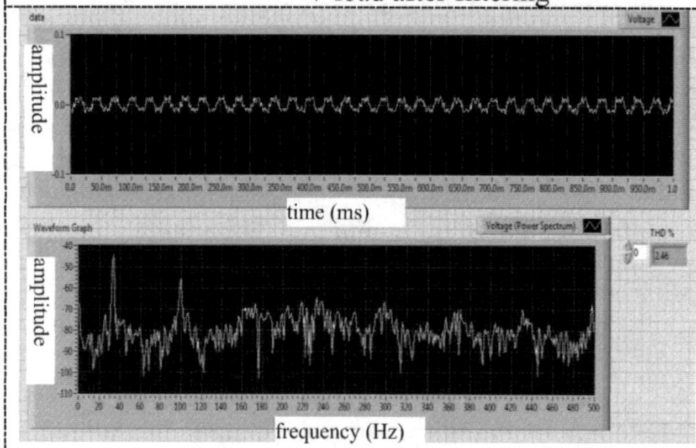

Figure 7. The current and frequency waveforms with a 3 Watt 220 V load after filtering

Fig. 6 shows the results of high order passive filter applications to reduce any harmonics. The figure indicates that sine waveform voltage has a period of 0.03 seconds with a fundamental frequency of 33 Hz with an amplitude of 6 dB. The harmonics on the 3rd (100 Hz), 5th (170 Hz), 7th (234 Hz), 9th (298 Hz), 11th (366 Hz), and 13th (435 Hz) have amplitude around -28 dB, -45 dB, -45 dB, -45 dB, -50 dB and -51 dB. It can be demonstrated that it appears that the 3rd order can be suppressed significantly by adding filters. The other order 9th, 11th, and 13th are also reduced compared with without filter. The total amplitude is around -27.6 dB with the THDv (%) is around 0.17 %.

The implementation of the filter can also be recognized in current measurements. Fig. 7 demonstrates sine waveform current, which period 0.03 seconds with a frequency of 33 Hz (fundamental) at an amplitude of -44 dB. In this figure, The harmonics on the 3rd (100 Hz), 5th (165 Hz), 7th (235 Hz), 9th (298 Hz), 11th (366 Hz), and 13th (435 Hz) order have amplitudes of around -57 dB, -68 dB, -63 dB, -66 dB, -70 dB, and -70 dB respectively. The total of the amplitude is around -58.9 dB with THDi (%) around 0.0003 %. These results demonstrate that the effect of the filter also significantly suppressed the amplitudes.

The example above shows the relationship between voltage and current before and after filtering using a single load. The summary relationship of the voltage and current can be presented in the following paragraph.

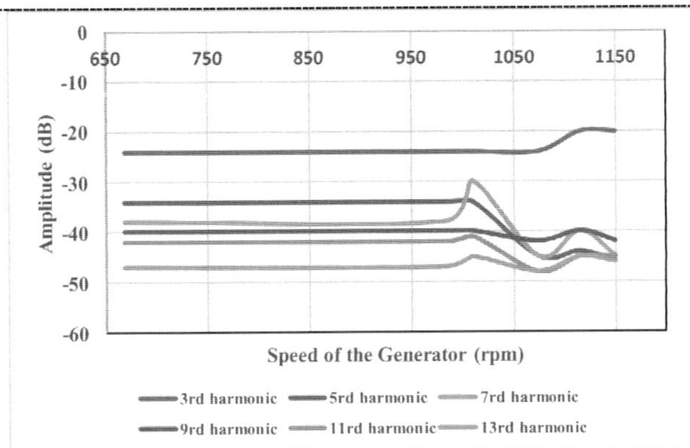

Figure 8. Voltage harmonics with various speed before filtering

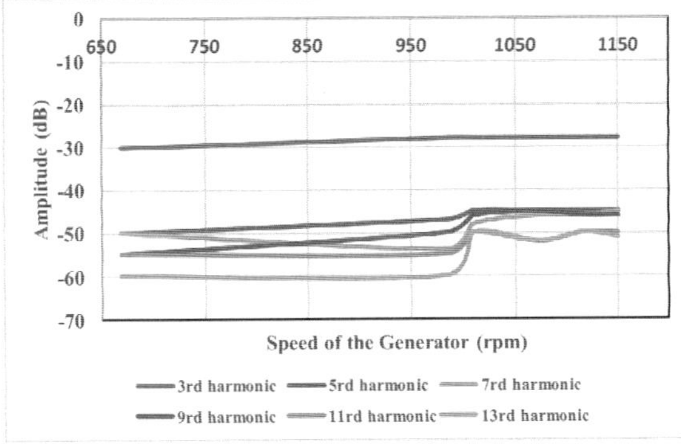

Figure 9. Voltage harmonics with various speed after filtering

Based on the two voltage harmonic graphs in Fig. 8 (before filtering) and Fig. 9 (after filtering), it is indicated that after filtering, the harmonic waveform peak decreases. The value capacitors and inductors as the harmonic dampers can also be investigated further to obtain better dampening. In addition, the magnitude of harmonics after filtering seems reasonably converge above 1000 rpm. It might be the characteristic of the magnetic field of the induction motor itself, which is much better at high speed, producing fewer harmonics.

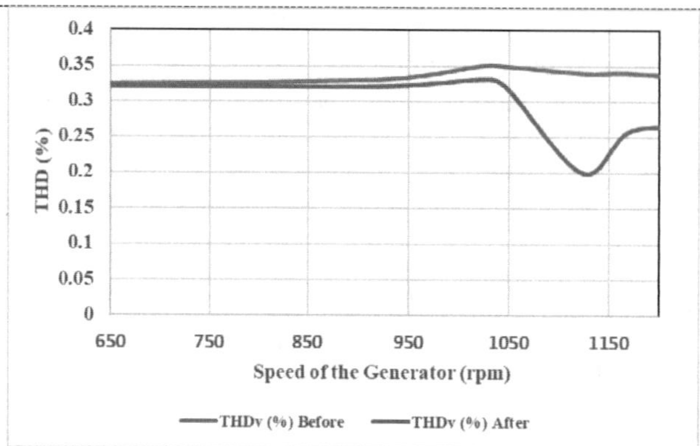

Figure 10. Summary of the total harmonic distortion before and after the filtering

Fig. 10 demonstrates the THD according to the variation of the generator speed. The THD decreased after filtering, especially above 1000 rpm, even though it is still below 5% as standardized. This graph also shows the practical application in reducing the harmonics of the voltage using simple filtering, especially for the application in the micro-hydro.

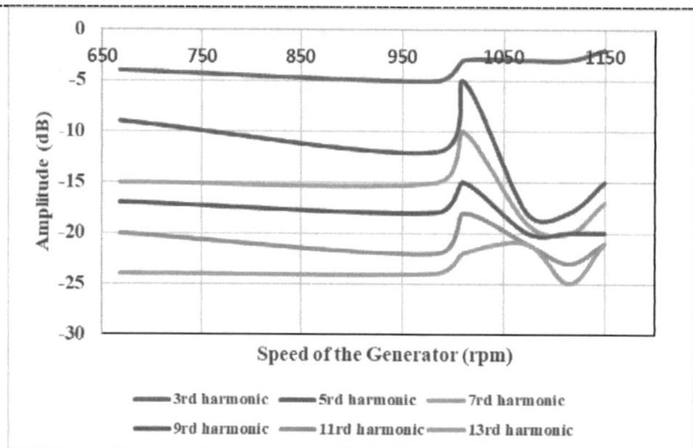

Figure 11. Current harmonics with various speed before filtering

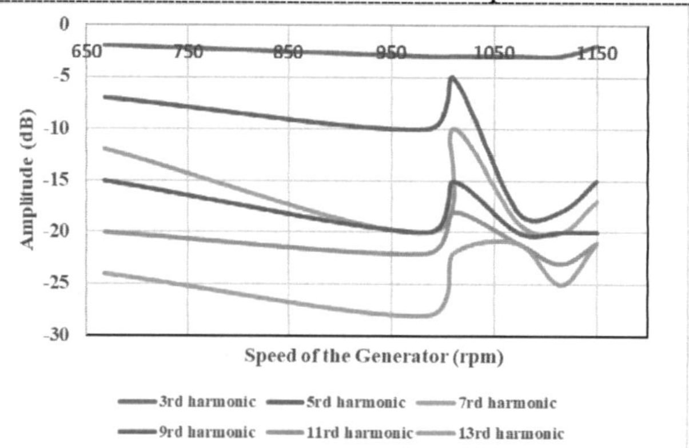

Figure 12. Current harmonics with various speed after filtering

Fig. 10 and 11 show the relationship of various harmonics amplitude under the various speed of the induction generator before and after the filtering. The current measurements also employ similar filter as given by voltage measurements. In the case of voltage harmonic wave graph, after filtering, the peak of the harmonic wave decreases by the bandpass filter. However, for the current measurement, the filter can seem more effective at low speed. In addition, between 950 and 1050, a small bump or increase occurs, which indicates the signature characteristic of the induction generator. In other words, the rpm from 1010 to 1150 clearly shows that the bandpass filter cannot filter or reduce the current harmonic waves. Further effort was carried out to try reducing these harmonics at these specific areas by applying other types of filters. The high order passive filter type was applied further to understand how the harmonic reduction can be realized. The results are given in Fig 13.

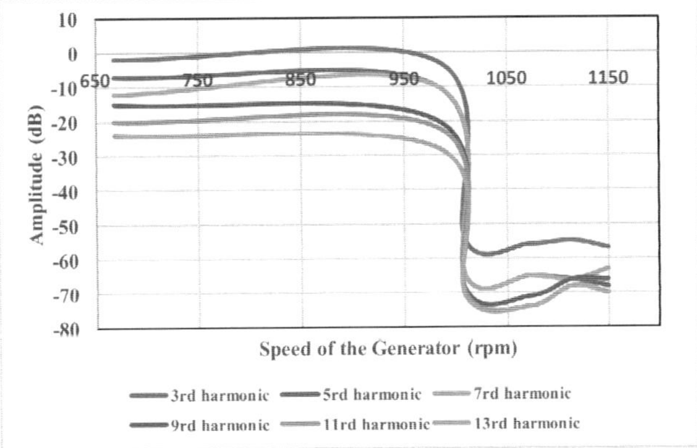

Figure 13. Current harmonics with various speed after filtering

Fig 13 indicates the applicability and functionality of the high pass order filter as one of the solutions to reduce the harmonics. This filter seems to show better performance at the high-speed generator. At the low speed, the higher harmonics can be reduced significantly, similar to the bandpass filter, but it is not for the 3rd harmonics. In addition, the high order filter also useful at the high speeds of the induction generator. This filter must also be applied carefully to avoid the reduction of fundamental current as well.

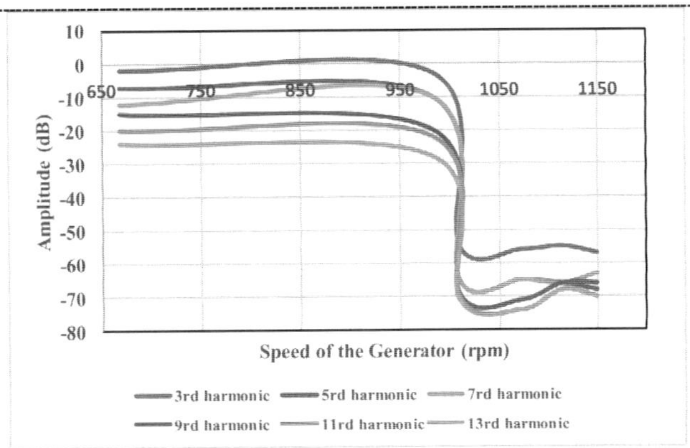

Figure 14. THDi before and after filtering

Fig. 14 shows that the THDi graph shows the comparison of the harmonic reductions between the Band Pass Filter and High Order Passive Filter. The measured harmonics using the bandpass filter are up to around 64%, compared with the high order passive filter, i.e., maximum for about 0.38%. This harmonic reduction provides a practical solution to reduce harmonics that are produced by the generator.

5. Conclusions

This paper demonstrates the application of filters performance for the mini pico or micro-hydro simulator applications. The paper presents the practical, low cost, and simple approaches of additional filters to the generator for better performance of the voltage and current output. This paper demonstrates the applicability of the simulator as well for the educational purposes. The Bandpass filter able to reduce voltage harmonics without load, which is THDv 0.33% to 0.26% even though it is still far in the IEEE standard i.e., 5%, and the bandpass filter has no effect while the system connected with the load. The high order passive filter can reduce the harmonic wave with THD 67 % becomes THDi 0.003% and THDv 0.40% to 0.17%. However, further investigations are still needed to ensure that the generated current is not excessively filtered. *Style and spacing*

Acknowledgment
We would like to acknowledge the Universitas Pamulang and Ristekdikti through the PDP scheme for the funding to carried out the research.

6. References

[1] Erinofiardi, P. Gokhale, A. Date et al., "A Review on Micro Hydropower in Indonesia," Energy Procedia, vol. 110, pp. 316-321, 2017/03/01/, 2017.
[2] Rimbawati, A. A. Hutasuhut, F. I. Pasaribu et al., "Design of motor induction 3-Phase from waste industry to generator for microhydro at isolated village," IOP Conference Series: Materials Science and Engineering, vol. 237, pp. 012021, 2017/09, 2017.
[3] Y. S. Syaiful Bakhri, "Design of Mini MicroHydro Simulator Based on Pelton Turbine (In Indonesia)," EPIC, vol. 2, no. 1, 2019.
[4] C. P. Ion, and C. Marinescu, "Autonomous micro hydro power plant with induction generator," Renewable Energy, vol. 36, no. 8, pp. 2259-2267, 2011.
[5] M. Shah, and S. Somkun, "Efficiency Evaluation of Three Phase and Single Phase C2C Self-Excited Induction Generator for Micro Hydro Power Application," Energy Procedia, vol. 138, pp. 193-198, 2017/10/01/, 2017.
[6] B. V. Nair, M. Mathew, A. Joseph et al., "Design of Micro-Hydro power plant using an induction motor as a generator." pp. 1-5.
[7] A. Mishra, P. M. Tripathi, and K. Chatterjee, "A review of harmonic elimination techniques in grid connected doubly fed induction generator based wind energy system," Renewable and Sustainable Energy Reviews, vol. 89, pp. 1-15, 2018.
[8] X. Liang, and Y. Luy, "Harmonic Analysis for Induction Motors." pp. 172-177.
[9] M. Nayeripour, and M. Mahdi Mansouri, "An advanced analytical calculation and modeling of the electrical and mechanical harmonics behavior of Doubly Fed Induction Generator in wind turbine," Renewable Energy, vol. 81, pp. 275-285, 2015.
[10] M. S. Muhammad Habibi Lubis, "Analysis of the Effect of Large Value of Excitation Capacitors on the Characteristics of Zero Load and Load on the Induction Motor as a Self-Strengthening Generator (MISG) (in Indonesia)," Singuda Ensikom, University of Sumatra Utara, vol. vol. 1, , no. 3, 2013.
[11] H. G. Fika Priliasari, "Study of the Effects of Harmonics on Electric Current on the Amount of Reduced Power Capacity (Kva) Installed Distribution Transformer, Case Study: Distribution Transformer of PT. PLN (Persero) Bekasi Raya region (in Indonesian)," Electrician, vol. 1 no. 1,, 2007.
[12] C. P. Jawahar, and P. A. Michael, "A review on turbines for micro hydro power plant," Renewable and Sustainable Energy Reviews, vol. 72, pp. 882-887, 2017.

IOP Conf. Series: Materials Science and Engineering **938** (2020) 012010 doi:10.1088/1757-899X/938/1/012010

SENTIPUBLIKO: Sentiment Analysis of *Repost Jejemon Messages* using Hybrid Approach Algorithm

Adomar L. Ilao and Arnel C. Fajardo

Technological Institute of the Philippines
Quezon City, Philippines

E-mail:alilao@mcl.edu.ph,acfajardo@gmail.com

Abstract. Jejemon language becomes a form of communication dialect. It was a form of expression used by a particular social group unknown as Jejemon. However, the Jejemon expression has different formats ranging from basic form of changing letter to number, lowercase letter to uppercase letter, inserting shortcut texts into more complicated format. This paper aims to classify Jejemon tweet whether it is a positive, negative or neutral sentiment through sentiment analysis techniques. Experiment included translation of Jejemon formatted tweet, reduction of sentiment scores on repost tweets and sentiment classification. Analysis of experiment results involves Paired T-Test, confusion matrix, precision, recall, f-score and accuracy. Evidently, translated Jejemon tweet resulted 78.5% similar from the actual message using cosine similarity algorithm. Furthermore, Paired T-Test shows no significant difference between new sentiment scores from translated expression and actual sentiment scores using Hybrid Algorithm. Sentiment analysis metrics such as precision, recall, f-score and accuracy show acceptable values of 71%, 76%, 71% and 73% respectively.

1. Introduction

Human language is constantly changing. It happens across time and social groups. These changes on human language yields negative perception from people who are unable to cope with new vocabulary or new visual representation. The downside of the new trends might result possible miscommunication between social groups.

In the Philippines, several languages were invented to cater specific social groups. Predominantly, millennial and member of the third sex are the most socially active group in terms of modern language expression [1]. They developed new language semantics such as Jejemon (p30pL3, o+h3r, pl4c3s) [2] and bikemon (Aglipay, Chiquito, Churchill). However, the most controversial language trend in 2016 is Jejemon [3].

Nevertheless, Jejemon language users suffered from weak speaking ability and mangled word spelling as observed by English teachers. They average 12 text messages everyday. Each jejemon message normally is composed of symbols and phonetics. However, the advent of technological communication medium focuses on social connection such as social media created new Jejemon followers in Twitter aside from JEJETYPING, wearing a jeje-hat and jeje-photos online [4].

The overwhelming success of Jejemon language lead to an award as Word of the Year in 2010 by the Filipinas Institute of Translation Incorporated based on significant impact on Filipino life in terms of socio-cultural, political, social and economic [5].

As an influential millennial language, Jejemon expression is a result of self-expression which designed to resolve concerns on limited available space provided by text messages and Twitter [5].

However, it is significantly important beyond translation to understand the actual feeling behind the person's message or opinion.

Classification of one's opinion can be done through sentiment analysis. It can classify opinion whether positive, negative or neutral via polarity score. Unfortunately, other factor to consider is the impact of repost known as retweeting in Twitter towards sentiment classification in a document level. This paper is designed to classify the sentiment of a phenomenon language developed by Jejemon generation. Its complexity to analyze the Jejemon expression using natural processing language is the key motivation of this paper. In classifying Jejemon tweets whether "positive sentiment" or "negative sentiment" relies on sentiment polarity value.

Hybrid algorithm developed by Ilao and Fajardo [6] will be used to determine sentiment polarity values that was proven better than SentiWordNet and VADER algorithm as tested to different political datasets. Furthermore, integration of string similarity algorithm in reducing sentiment polarity score for repost or retweet messages will provide better understanding of the side effect of repost messages in a document level sentiment evaluation.

The paper is organized by sections. Section 1 elaborated Jejemon language origin, sentiment analysis and possible issue on repost message. Section 2 enumerated vital concepts related to the paper. Section 3 discusses the detailed stages of sentiment analysis. Section 4 described characteristics of collected instances. Section 5 cited statistical instrument and sentiment analysis measurement to provided performance analysis. Section 6 explained performance and evaluation of translation, dataset and algorithms. Lastly, Section 7 presented the general conclusion of the study.

2. Review of Related Literatures

2.1. Jejemon Word Structure

Jejemon language became popular in 2010 based on the limited space available for text messages and tweets [5]. It is primarily composed of alphabet known as Jejebet.

Jejebet uses Roman alphabet (a-z), Arabic numerals (0-9) and other special characters (@, !, ~, ?, .). However, Jejemon word is more than jejebet alphabet. It has several types of pattern namely alternating capitalization, over-usage of the letters H, X or Z and mixture of numeric characters and English alphabet [5] as shown in Table 1.

Table 1. Example of Jejemon Expression [5]

Characteristics of Jejemon	Example
Insertion of unnecessary numbers and letters.	phfue or p0w
Unique orthography based on how the words sound	eHyUoeW fPuoEh
Unconventional use of punctuations	psenxa na ha!!
Numbers to substitute letters	bzt4h
Alternate use of lower and upper case	WE wnT 2 BE~ P0wh.
Use of onomatopoiec lexis/emotional language	tnx pfowh jejeje
Lengthening of vowels and consonants	TAMAAA!
Substitution of spelling	M@q

2.2. Jejemon Translation

Jejemon translation is available online. One particular online translator is Jejemon Translator found in http://173.254.110.65/jejeschool/index.php. It is capable to translate English language to Jejemon expression as shown in Figure 1.

IOP Conf. Series: Materials Science and Engineering **938** (2020) 012010 doi:10.1088/1757-899X/938/1/012010

Figure 1. English Language to Jejemon Language Translation.

It was applied to transform the English thesis document version [7] to jejenese or Jejemon language document version available in iskwiki.upd.edu.ph. It shows the reliability of the Jejemon translator as an effective technical tool. The jejemon document revision demonstrated combination of three Jejemon techniques namely numbers to substitute letters, symbols to substitute letters and alternate use of lowercase and uppercase letters.

An alternative online Jejemon translator can be found in http://akosijairah.blogspot.com/2010/04/Jejemon-translator-v3_28.html as shown in Figure 2. The translated expression is comprised of case conversion, p0wh insertion and modification of word to a totally different spelling.

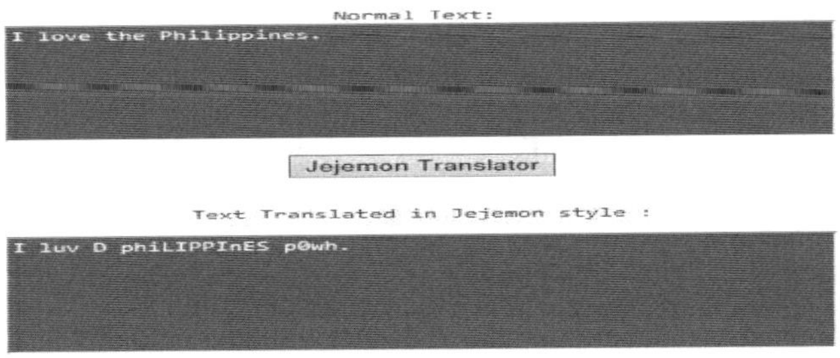

Figure 2. English Language to Jejemon Language Translation.

2.3. Dictionary Substitution Approach

Dictionary Substitution Approach is a technique commonly identified as search and replace approach. The key characteristic is to match word from the corpus. However, when multiple entries are found from the corpus, randomize word will be selected from list possible alternative [8].

Jejemon language does not follow specific pattern. Through Dictionary substitution approach it can replace non-standard Jejemon token into a meaningful context of English or Filipino word. English or Filipino sentence found in Table 2 was translated to Jejemon equivalence from three sources. It shows three different techniques of Jejemon translation.

Table 2. Sample Online English Expression Translated into Jejemon Expression

English Sentence	URL Source of Jejemon Translator	Jejemon Expression
I would like to know more about you, care to tell me your name? Hehehehe!	https://pinoychronicle.wordpress.com/Jejemon/	i wuD LLyK tO knOw moR3 bOut u. crE 2 t3ll mE yur N@me? jejejejeje!
	Online Jejemon Translator (http://173.254.110.65/jejeschool/index.php)	1 wUD 77yk +0 kn0w mUhr3 4b0U+ U', cr +0 +377 m3 U'r nm3?" j3j3j3h3!
	Online Jejemon Translator (http://akosijairah.blogspot.com/2010/04/Jejemon-translator-v3_28.html)	i WOUlD Lyk To KNow more aboutz u, CeyRTo Tell me uR nMe, n0H? JEjEJEjE LOLz!

2.4. Lexicon-Based Algorithm

Lexicon-based Algorithm works by defining rules to classify the opinion which is created by tokenizing every sentence in each document and testing if the token or word is present in the database [8]. It is based on rule which is composed of antecedent and consequent. An antecedent defines a condition and consists of either a token or a sequence of tokens. This process provides a technique to single out positive, negative or neutral about the subjective opinion [9][10][11].

It uses sentiment lexicon to assign a polarity value. A lexicon is comprised of words or phrase where each label is categorized based on polarity value whether positive or negative orientation [12]. In building a sentiment lexicon have three strategies namely hand-craft elaboration, automatic expansion from an initial list of seed words and corpus-based approach.

A comparative study [13] on Lexicon-based review involving AFINN, General Inquirer, Micro-WNOP, Opinion Lexicon, SentiSense, SentiWordNet, Subjectivity Lexicon and WordNet-Affect. The investigation resulted 78% accuracy towards SentiWordNet which utilizes WordNet corpus

2.5. String Similarity Algorithm

String based similarity measurement defines the similarity of strings in terms of the longest prefix common to both strings. It is applied to several fields namely data cleaning, data integration, error checking or pattern recognition [14]. It uses either character-based or term-based technique [15]. The commonly used string based similarity measurement. It is a term-based technique known as edit distance algorithm which performs minimum number of insertions, deletions or substitutions to string1 to string 2 [14].

The comparative study on edit distance algorithms namely Q-gram similarity, cosine similarity and dice coefficient similarity. The study resulted in favor of cosine similarity algorithm with an average accuracy of 63% [16]. Cosine similarity algorithm measures two-finite-dimensional vectors of the same dimension [15]. Furthermore, cosine similarity was applied to analyze the similarity of sentiment scores from SentiWordNet and the similarity of each sentiment score contributed to product review rating prediction [17].

2.6. Hybrid Polarity Score Algorithm

Each synset polarity score is derived by computing the average of SentiWordNet algorithm and VADER algorithm as elaborated in Equation 1, Equation 2 and Equation 3 [18]. Normally, VADER polarity score use compound score to determine the sentiment classification. However, Ilao's Hybrid algorithm used VADER's positive and negative scores to derived sentiment score.

Hybrid positive score = (SentiWordNet Positive Score + VADER Positive Score)/2 (1)

Hybrid negative score =(SentiWordNet Negative Score + VADER Positive Score))/2 (2)

Over-all Hybrid Score (OHS) = (Hybrid positive score – Hybrid negative Score) (3)

Where:

If OHS >0, then sentiment is "Positive".

If OHS<0, then sentiment is "Negative".

If OHS=0, then sentiment is "Neutral".

The number of occurrences of positive, negative and neutral tweets will determine the over-all sentiment of the entire population of collected political tweets. The hybrid approach was experimented to different political datasets. The experiment yielded 88.33% accuracy better than SentiWordNet and VADER algorithm.

3. Methodology

The study was designed to implement sentiment analysis approach as illustrated in Figure 3.

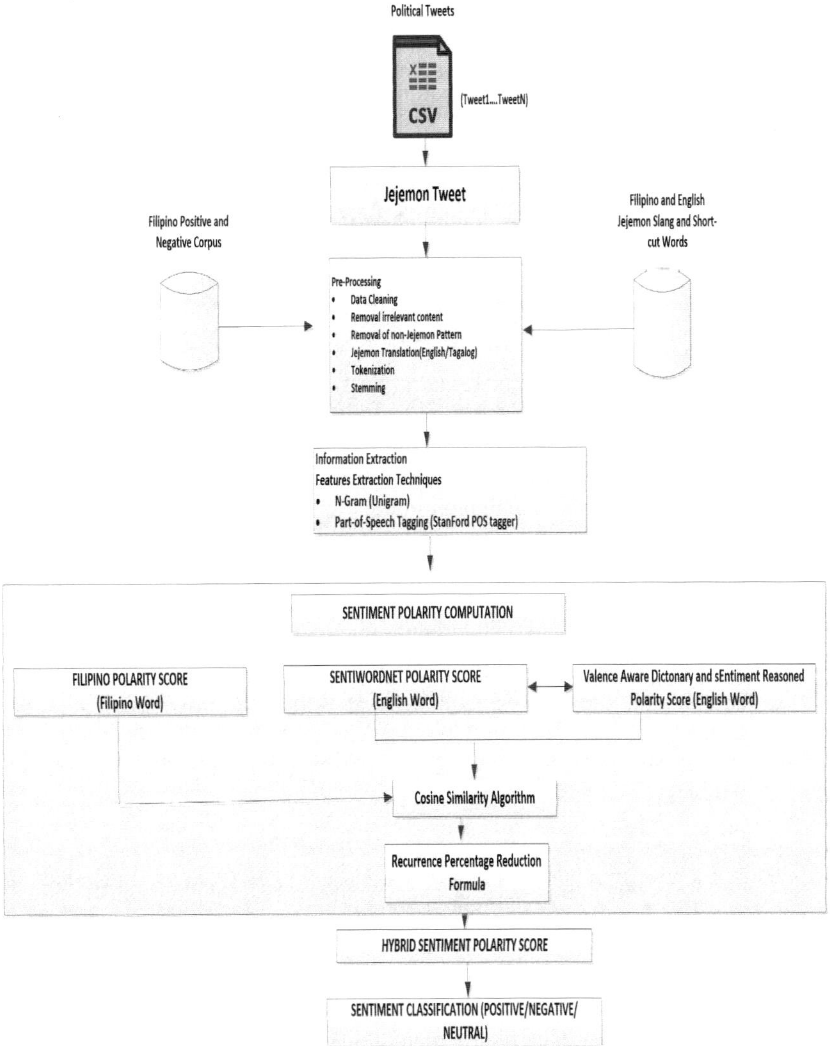

Figure 3. Jejemon Repost Hybird Polarity Score Algorithm Conceptual Framework

It involves several stages such as data cleaning, removal of irrelevant words, dictionary substitution and loan translation approach, tokenization, stemming, feature extraction, hybrid sentiment polarity score approach (SentiWordNet Score, VADER score, Filipino Score, Recurrence Percentage Reduction Score), and sentiment classification.

3.1. Pre-Processing

3.1.1. Data Cleaning

It involves elimination of unwanted expression such as removal of punctuations, website URL, emoticons, special characters and apostrophes.

3.1.2. Removal of Irrelevant Words

It is intended to remove irrelevant contents such as slang words [19], stop words, Jejemon expression (example: poWH, jejeje, xD) and non-Jejemon formatted text.

3.1.3. Dictionary Substitution and Loan Translation Approach

Jejemon lexeme might contains abbreviation or shortcut text in English or Filipino expression. A CSV file and Text file will stored shortcut English text, shortcut Filipino text and Jejemon most sentimental words. When translated lexeme is found from the seedlist, detected text will be replaced into its actual value. It also involves translation of Jejemon expression to purely English expression, purely Filipino expression and combination of both languages.

Translation utilized enchant libraries [20] and collection of Filipino words from Tagalog dictionary [21] to cross-checked spelling or possible word suggestion in transforming complicated Jejemon expression.

Furthermore, customized Tagalog dictionary was constructed from collection of rubbish words of English or Filipino words will resolve some ambiguity coming from Jejemon expression translation.

3.1.4. Tokenization

The process of tokenization segments where Jejemon expression might be in the form of paragraph, sentences and word into a lexeme (single word).

3.1.5. Stemming

In this stage, Jejemon lexeme will subjected to extraction of base word by simplifying plural form to singular or by removing prefix, suffix and infix.

3.2. Feature Extraction

3.2.1. N-gram Approach

The study will implement a hybrid approach in classifying the given Jejemon expression whether positive, negative and neutral sentiment. The two lexicon-based algorithms that was proven effective by Ilao's study in 2019 combined SentiWordNet and VADER algorithm polarity scores as shown in Equation 1, Equation 2 and Equation 3 in which uni-gram and tri-gram approach were implemented by the said algorithms respectively.

3.2.2. Part-of-Speech (POS) Tagging

Each English lexeme will be tagged via StanFord POS Tagger. Each tag will determine whether the lexeme can be a source of sentiment of SentiWordNet.

However, Filipino lexeme will not undergo POS Tagging procedure. The study will used Tagalog Corpus in identifying positive and negative political words.

3.3. Hybrid Approach

3.3.1. SentiWordNet Polarity Score

SentiWordNet polarity score features positive, negative and neutral scores. Nonetheless, Hybrid approach requires positive and negative elements of its derived polarity score.

3.3.2. VADER Polarity Score

Classification of sentiment using VADER normally uses compound score. However, hybrid approach will apply positive score and negative score as shown in Equation 1.0 and Equation 2.0.

3.3.3. Filipino Sentiment Polarity Score

Filipino sentiment polarity score will be applied for Filipino expression based on Equation 4.0. For every detected positive or negative word from collected list of commonly used tagalog political sentiment word will be scored as 1 point.

Filipino Polarity Score (FPS) = (Number of Positive Words-Number of Negative Words)/Number of Words in the Tweet (4)

Where:

If FOSS >0, then sentiment is "Positive".
If FOSS<0, then sentiment is "Negative".
If FOSS =0, then sentiment is "Neutral".

If translated expression is 85% classified as an "English Expression", Over-all Sentiment Polarity Score will be computed by Equation 1, Equation 2 and Equation 3. If translated expression is 85% classified as "Filipino Expression", Over-all Sentiment Polarity Score will be derived by Equation 4. Otherwise, Over-all Sentiment Polarity Score will be calculated by Equation 5.0.

Average Polarity Score (APS) = (OHS+FPS)/2 (5)

Where:

If APS>0, then sentiment is "Positive".
If APS <0, then sentiment is "Negative".
If APS =0, then sentiment is "Neutral".

3.3.4. Cosine Similarity

Cosine Similarity algorithm will be used to determine if there are some similarities between posts. If cosine similarity returns a string similarity value of 70%, there exists a significantly identical tweet between the lists of collected tweets which is considered as a repost message.

3.3.5. Recurrence Percentage Reduction Score

After string similarity algorithm identified similarity between tweets, Equation 6.0 and Equation 7.0 will define reduction score from previously derived value and generate a new polarity score for repost message.

Relative Frequency Rate = (number of item repost /total number of post messages)x100 (6)

Recurrence Percentage Reduction Score= Over-all Hybrid Score - (Relative Frequency Rate x Over-all Hybrid Score) (7)

3.3.6. Sentiment Classification

The study was designed to perform sentiment analysis in a document level. Each Jejemon expression will be classified whether positive, negative or neutral sentiment. The classification is dependent on the Over-Hybrid Score (OHS), Filipino Polarity Score (FPS) or Average Polarity Score (APS) as described by case scenario listed below.

Case *1*: Jejemon Expression translated as English Expression
If OHS >0, then sentiment is "Positive".
If OHS<0, then sentiment is "Negative".
If OHS=0, then sentiment is "Neutral".

Case *2*: Jejemon Expression translated as Filipino Expression
If FOSS >0, then sentiment is "Positive".
If FOSS<0, then sentiment is "Negative".

If FOSS =0, then sentiment is "Neutral".
Case *3*: Jejemon Expression after translation falls below 85% classified as Filipino or English Expression
If APS>0, then sentiment is "Positive".
If APS <0, then sentiment is "Negative".
If APS =0, then sentiment is "Neutral".

Afterward classification, each identified sentiment will be counted individually; the highest number of frequencies between positive, negative or neutral will generally determine the over-all sentiment classification of the collected instances.

4. Data Source

The dataset was collected from Twitter accounts namely Jejemonilao and Jejemonkami from January 15,2020 up to February 29,2020 as stated in Table 3, Table IV, Table V and Table VI with 131 instances.

Tweets have a maximum number of 37 tokens, minimum number of 3 tokens and an average of 12 tokens per instance.

Jejemonilao tweets followed Jejemon format available in http://akosijairah.blogspot.com/2010/04/Jejemon-translator-v3_28.html considered as controlled environment. However, uncontrolled environment instances were extracted from Jejemonkami tweets. Tweets were formatted based on preference of participants from any available Jejemon patterns. Moreover, collected dataset is comprised of purely English Jejemon tweet, purely Filipino Jejemon tweet and combination of both.

Table 3. Sample of Jejemon Format Domain

Original Expression	Jejemon Format Domain	Expression	Frequencies
Government plays a good job during Taal Eruption.	Online Resource (Controlled Environment)	G0V3rNm3nt pLaYZ a G00D J0B DuRiNg TaAl ErUpTi0n.	39
	Personal Preference (Uncontrolled Environment)	goVerNmEnt pLyz a goOD job duRIng Taal eRUPtioN p0WH.	92

Table 4. Tweet Post Frequencies

Tweet Post Tweets	Frequencies
Single Post	116
Repost	15

Table 5. Jejemon Collected Tweet Used Language

Jejemon Tweet used Languages	Frequencies	No. Instances under Controlled Environment	No. Instances under Uncontrolled Environment
English	113	35	78
Filipino	14	1	11
Combination	4	3	3

Table 6. Sentiment Classified Frequencies

Sentiment Classification	Frequencies
Positive	29
Negative	76
Neutral	26

5. Measurement of Algorithm Performance

Algorithm performance will be validated by means of paired T-Test, accuracy, recall, f-score and precision using Equation 8. Equation 9, Equation 10 and Equation 11 via confusion matrix values as shown in Table 7.

Table 7. Confusion Matrix

		Predicted	
		Negative (F)	Positive (T)
Actual	Negative (F)	FF	FT
	Positive (T)	TF	TT

$$\text{Accuracy} = \frac{FF+TT}{FF+FT+TF+TT} \quad (8)$$

$$\text{Precision} = \frac{TT}{FT+TT} \quad (9)$$

$$\text{Recall}=\frac{TT}{TT+TF} \quad (10)$$

$$\text{F-Score}=\frac{2*Precision*Recall}{Precision+Recall} \quad (11)$$

Where:
FF: the frequency of correctly predicted negative emotion.
FT: the frequency of incorrect predicted negative emotion.
TF: the frequency of incorrect predicted positive emotion
TT: the frequency of correct predicted positive emotion.

6. Experiment

The study evaluated the translation success rate of Jejemon instances to counterpart language via string similarity algorithm presented in Table 8. Each instance from each domain is compared to the actual expression.

Similarity Percentage is based cosine similarity value of translated expression against actual expression. Similarity values fall from 70% and above were classified as "Strong Similarity" otherwise considered as "Weak Similarity".

Table 8. Similarity Percentage of Translated Expression and Actual Expression

Jejemon Format Domain	**No. of Strong Similarity (70% and Above)**	**No. of Weak Similarity (Below 70%)**	**Average Cosine Similarity Percentage**
Online Resource (Controlled Environment)	92%	7%	81%
Personal Preference (Uncontrolled Environment)	73%	27%	76%

Table 9 shows the lowest and highest similarity percentage of the entire dataset per language. The controlled group's translation similarity percentage ranges from 23% up to 100%. The factors which affect translation success rate are word case, missing letters that were omitted by Jejemon expression when translated lead to a different English or Filipino expression due to ambiguity, proper space between Jejemon expression and single/plural forms. However, uncontrolled group's similarity percentage ranges 24% up to 100%. Significant difference from controlled group, uncontrolled group's factors are special characters such as Ë, ë, $, @, or ü ; repetition of letters namely "z","s"; two representation of 8 either "te" or "ate" , "i" either "!" or "1", "a" either "4" or "@" and Filipino word interchangeably used "po" or "poh".

Table 9. Cosine Similarity Percentages per Language

Jejemon Format Domain	Languages	Lowest Similarity Percentages	Highest Similarity Percentages	Average of Similarity Percentage
Online Resource (Controlled Environment)	English	23%	100%	81%
	Filipino	89%	89%	89%
	Combination	67%	89%	74%
Personal Preference (Uncontrolled Environment)	English	25%	100%	78%
	Filipino	24%	100%	39%
	Combination	25%	70%	55%
Average		42%	91%	69%

Furthermore, Table 9 presented the lowest similarity percentage is 23% under English language of Controlled Environment. The event was brought by several wrongly translated words due to ambiguity. Some Jejemon tweets have missing letters, translation of the given scenario relies on selection from a given list of suggested words. Other scenario, some words should be express in capital letter.

While English and Filipino expression (combination) of Uncontrolled Environment gained the lowest value of 70% under highest similarity percentage. Uncontrolled Environment encountered similar condition from Controlled Environment where missing letters might be rubbish word or an abbreviation but it cannot be disregarded. For example, "plyz" was replaced by "plys" based on the suggested words from English or Tagalog Dictionary where it should the word "plays". There are scenarios where an English word was taken as a Filipino word or vice versa.

After Jejemon translation, the new dataset comprised of English, Filipino or combination underwent pre-selection. As stated on previous studies on machine translation on several Filipino dialects, accuracy rates are 70.67% [22] and 69.5% [23]. Pre-selection of instances will be based on similarity percentage between 70% up to 100%. Qualified instances are revealed in Table 10.

Table 10. New Instances after Pre-Selection

Languages	Controlled Environment	Uncontrolled Environment
English	33	64
Filipino	0	2
Combination	1	1
Total	34	67

Table 10 shows 12% decreased on the number of instances under the controlled environment from 39 instances down to 34 instances. Similarly, 27% decreased on uncontrolled environment instances from 92 instances down to 67 instances. It also shows the most retained translated tweet is English tweet with 94% retention. While Filipino tweets suffered the most reduction with 18% retention out of 11 instances.

IOP Conf. Series: Materials Science and Engineering **938** (2020) 012010 doi:10.1088/1757-899X/938/1/012010

Table 11. Paired T-Test of Computed Sentiment Scores

		Mean	N	Mean Difference	p-value	Interpretation
Uncontrolled	New Score	-.073824	67	-.002619	.821	Not Significant
	Original Score	-.071204	67			
Controlled	New Score	-.062557	34	.016663	.257	Not Significant
	Original Score	-.079220	34			
Similarity Reduction	New Score	-.163912	32	-.017796	.246	Not Significant
	Original Score	-.146115	32			

Table 11 described the relationship between different sentiments' scores namely new score is derived from translated expression, original score is derived from the actual expression. Furthermore, cosine similarity found instances with 70% similarity considered as repost messages. All repost sentiments' scores underwent either 1% or 3% reduction. As stated in Table 11, p-value results are 0.821, 0.257 and 0.246. They are greater than 0.05 which implied there is no significant difference between new scores and original scores under controlled environment, uncontrolled environment and after application of recurrence percentage reduction scores.

Table 12. Confusion Matrix of Controlled Environment

		Predicted	
		Negative (F)	**Positive (T)**
Actual	**Negative (F)**	11	1
	Positive (T)	7	6

Table 13. Confusion Matrix of Uncontrolled Environment

		Predicted	
		Negative (F)	**Positive (T)**
Actual	**Negative (F)**	31	3
	Positive (T)	9	11

Table 12 and Table 13 shown number of positive and negatives instances classified correctly or incorrectly. It appears 59 instances out of 79 instances were classified correctly and 20 instances were classified incorrectly.

Table 14. Sentiment Analysis Metric Summary

Jejemon Format Domain	**Label**	**Precision**	**Recall**	**F-Score**	**Accuracy**
Online Resource (Controlled Environment)	Positive	46%	86%	60%	68%
	Negative	92%	61%	73%	
Personal Preference (Uncontrolled Environment)	Positive	55%	79%	65%	78%
	Negative	91%	78%	81%	
Average		71%	76%	71%	73%

Table 14 shows both negative sentiments resulted high precision with at least 91% and lowest precision value is 46% under Positive Controlled Environment. While, recall highest value is 86% falls under positive sentiment of Controlled Environment Domain; however the rest of the recall values are 61% up to 79%. Lastly, the highest f-score value is 84% under the negative sentiments of Uncontrolled Environment Domain whereas the remaining values are within 60% up to 84%.

The hybrid model provided accuracies of 68% (Controlled Environment Domain), 79% (Uncontrolled Environment Domain) and an average of 74% from collected Jejemon format domain. Generally, average scores of precision, recall, f-score and accuracy resulted 70% and above in term of classification performance towards the different Jejemon format domains.

7. Conclusion

This paper experimented on sentiment analysis solely focus on Filipino fascination in expressing their idea through Jejemon language.

However, Jejemon language raised issues on millennial English proficiency. Jejemon language as a form of communication is normally expressed through short-text and special representation of word or expression. Furthermore, Jejemon expression offered several language structure ranging simple to complicated techniques.

In general, the study of machine translation from Jejemon expression into counterpart language namely English, Filipino or combination (English and Filipino) yielded successful conversion based on the cosine similarity done. The similarity rates of translated expression against actual expression gained 76% and 81% for controlled and uncontrolled environment.

Even though some instances were not converted exactly as compared to actual expression, paired T-Test shows no significant difference between original polarity score and new polarity score derived from the two dataset domains. Moreover, repost messages that underwent reduction through recurrence percentage reduction technique do not shows significant difference from the two domains based on the available data.

Sentiment classification through hybrid approach resulted positively in terms of consistency and accuracy based on the average precision, recall and accuracy such as 71%, 76% and 73%. Average F-score of 71% signifies the acceptable classification performance since there is an uneven representation of sentiment instances from collected domains.

Acknowledgments
Authors would like to express their greatest gratitude to Graduate School of Technological Institute of the Philippines, Quezon City for the support of realization of this paper publication.

References

[1] Ramos, J. J. R. (2018). Bekimon in Social Networking Site: Modernized Idiolect and Sociolect. *International Journal of Social Sciences & Humanities*, *3*(2), 16-25.

[2] Nocon, N., Cuevas, G., Magat, D., Suministrado, P., & Cheng, C. (2014, October). NormAPI: An API for normalizing Filipino shortcut texts. In *2014 International Conference on Asian Language Processing (IALP)* (pp. 207-210). IEEE.

[3] Mongaya, K.M. (2010). (http://karlomongaya.wordpress.com/2010/07/09/the-internet-as -corrective-of-individualist-culture/)

[4] The Social and Educational Influences of Jejemon Texting Style.(2010) Date Retrieved: November 10, 2019

[5] Tubac, Angelo. (2017). LINGUISTIC INNOVATIONS IN THE JEJEMON PHENOMENON. 10.13140/RG.2.2.18434.07360.

[6] Ilao, A. & Fajardo, A. Sentiment Analysis of Tweet Messages using Hybrid Approach Algorithm. In 2019, The 17th International Conference on ICT and Knowledge Engineering (ICT-KE). IEEE.

[7] Cataan, J. C. (2011). *A 'World' Within the World: The Jejemons as the 'Other' Culture*, Unpublished Undergraduate Thesis, University of the Philippines College of Mass Communication.

[8] Raghunathan, K., & Krawczyk, S. 2009. CS224N: Investigating SMS Text Normalization using Statistical Machine Translation. Department of Computer Science, Stanford University.

[9] Devika, M. D., Sunitha, C., & Ganesh, A. (2016). Sentiment Analysis: A Comparative Study on Different Approaches. *Procedia Computer Science*, *87*, 44-49.

[10] Romanyshyn, M. (2013). Rule-Based Sentiment Analysis of Ukrainian Reviews. *International Journal of Artificial Intelligence & Applications*, *4*(4), 103

[11] Kawathekar, S. A., & Kshirsagar, M. M. (2012). Sentiments Analysis Using Hybrid Approach Involving Rule-Based & Support Vector Machines Methods. *IOSRJEN*, *2*(1), 55-58.

[12] Almatarneh S, Gamallo P (2018) A Lexicon Based Method to Search for Extreme Opinions. Plos One 13(5): E0197816. Https://DOI.Org/ 10.1371/Journal.Pone.0197816

[13] Cho, H., Lee, J. S., & Kim, S. (2013). Enhancing Lexicon-Based Review Classification by Merging And Revising Sentiment Dictionaries. In *Proceedings of the Sixth International Joint Conference on Natural Language Processing* (pp. 463-470).

[14] Lu, W., Du, X., Hadjieleftheriou, M., & Ooi, B. C. (2014). Efficiently Supporting Edit Distance Based String Similarity Search Using B+-Trees. *Ieee Transactions on Knowledge and Data Engineering*, *26*(12), 2983-2996.

[15] Khuat, Tung & Duc Hung, Nguyen & Thi My Hanh, Le. (2015). A Comparison of Algorithms used to Measure The Similarity Between Two Documents. International Journal of Advanced Research in Computer Engineering & Technology (IJARCET). 4. 1117-1121.

[16] Chen, H. (2012). *String Metric and Word Similarity Applied to Information Retrieval* (Doctoral Dissertation, Master's Thesis, School Of Computing. University Of Eastern Findland).

[17] Prasanna Lakshmi, K., Shraddha, V., Abhinava, V., Kavya, K., & Gayathri, R. (2017). Sentiment Analysis And Prediction Using Text Mining. Indian Journal Of Science And Technology, 10 (28). Doi: 10.17485/Ijst/2017/V10i28/113441

[18] Ilao, A. & Fajardo, A (2019). SENTIMENT ANALYSIS OF TWEET MESSAGES USING HYBRID APPROACH ALGORITHM. 17th International IEEE Conference of ICT and Knowledge Engineering 2019

[19] https://www.internetslang.com/trending.asp. Accessed Date : March 10,2020

[20] https://abiword.github.io/enchant/. Accessed Date : March 12,2020.

[21] https://github.com/raymelon/tagalog-dictionary-scraper/blob/master/tagalog_dict.txt. Accessed Date: March 12,2020.

[22] Domingo and R. Roxas. 2006. Utilizing Clues in Syntactic Relationships for Automatic Target Word Sense Disambiguation. Journal of Research for Science, Computing and Engineering. 3(3), 18-24.

[23] Alcantara and A. Borra. 2008. Constituent Structure for Filipino: Induction through Probabilistic Approaches. Proceedings of the 22nd Pacific Asia Conference on Language, Information and Computation (PACLIC). 113-122.

AUTHOR INDEX